THE SPREAD
OF THE SCIENTIFIC REVOLUTION
IN THE EUROPEAN PERIPHERY,
LATIN AMERICA AND EAST ASIA

DE DIVERSIS ARTIBUS

COLLECTION DE TRAVAUX
DE L'ACADÉMIE INTERNATIONALE
D'HISTOIRE DES SCIENCES

COLLECTION OF STUDIES
FROM THE INTERNATIONAL ACADEMY
OF THE HISTORY OF SCIENCE

DIRECTION
EDITORS

EMMANUEL
POULLE

ROBERT
HALLEUX

TOME 45 (N.S. 8)

BREPOLS

PROCEEDINGS OF THE XXth INTERNATIONAL CONGRESS
OF HISTORY OF SCIENCE (Liège, 20-26 July 1997)

VOLUME V

THE SPREAD
OF THE SCIENTIFIC REVOLUTION
IN THE EUROPEAN PERIPHERY,
LATIN AMERICA AND EAST ASIA

Edited by

Celina A. LÉRTORA MENDOZA, Efthymios NICOLAÏDIS

and Jan VANDERSMISSEN

BREPOLS

The XXth International Congress of History of Science was organized by the Belgian National Committee for Logic, History and Philosophy of Science with the support of :

D/2000/0099/10
ISBN 2-503-50889-8
Printed in the E.U. on acid-free paper

TABLE OF CONTENTS

AVANT-PROPOS

Efthymios NICOLAÏDIS

L'historiographie des sciences au XXᵉ siècle est marquée par l'étude de la " science classique ", c'est-à-dire par l'étude de la naissance et du développement des connaissances de la " Révolution scientifique ". Ce phénomène qui fut européen et succéda à la Renaissance européenne, deviendra un véritable courant de pensée qui sera au cours des siècles suivants diffusé dans toutes les parties du monde. A nos jours, la science est enseignée presque de la même manière dans les universités de la Terre et ses fondements se trouvent justement dans cette science européenne, devenue " science classique " pour tous.

L'étude de cette épopée du savoir humain continue et continuera d'occuper les historiens des sciences. Mais tandis que jusqu'alors — et pour cause — l'accent fut mis sur l'étude de la constitution même de ce savoir, peu de chercheurs se sont penchés sur sa diffusion dans les pays qui n'ont pas ou peu participé à son élaboration. Pourtant, l'introduction de ces connaissances a bouleversé bien des choses dans différentes parties du monde ; elle fut la condition *sine qua non* de la modernisation des pays de la périphérie pendant le XIXᵉ siècle.

Ainsi, ces dernières années, un intérêt s'est manifesté dans différents pays pour l'étude de la diffusion sur leur territoire du savoir de la science européenne de l'époque moderne. Plusieurs questions, d'un ordre nouveau pour l'histoire des sciences, se sont posées. Quels furent les mécanismes de la transmission du savoir vers les pays de la périphérie ? Comment ce savoir a-t-il été accepté ? Quelles furent les continuités et quelles furent les ruptures avec le savoir scientifique existant jusqu'alors dans ces pays ? Tentant d'y répondre, des équipes de chercheurs se sont formées, des travaux ont commencé à voir le jour et des collaborations internationales à se dessiner.

Dès alors, l'idée avait mûri d'organiser un Symposium sur ce thème au sein du XXᵉ Congrès International d'Histoire des Sciences. Ce Symposium aurait comme but de présenter un éventail de recherches actuelles sur le sujet, concernant les pays se trouvant à la périphérie de l'Europe (péninsule ibérique,

Balkans, Russie, pays scandinaves). Il se trouva que nos collègues sud-améri-
cains avaient tenté la même approche pour l'Amérique latine et que la diffusion de la science européenne en Chine est l'objet d'une activité historio-
graphique. Ainsi, la synthèse a été faite au présent Symposium, où les trois
unités présentées (périphérie européenne, Amérique latine et Chine) tentent de
couvrir un espace géographique varié et représentatif.

PART ONE

EUROPE

Bottlenecks : 18th Century Science and The Nation State

Arne Hessenbruch

In the 18th century, the town gate was a place where measurement was carried out routinely. Illustration I depicts a queue of carts bringing goods to market. They are lined up in front of a small house : the weighbridge. There is a line of soldiers, representatives of authority, the presence of whom guarantees that no one gets past without paying. The city wall in the background funnels the traffic so that all carts have to get past this or a similar point. The main places where indirect taxes were levied all had this funnelling effect in common : town gates, harbours, shops, mills, and turnpikes. All these places are in effect bottlenecks. Circulation flows through the bottleneck. The traders bring their merchandise to the bottleneck because they have to, for instance in order to get to the city market (thus " bottleneck " is a much more appropriate concept than, say, " junction "). It is not only merchandise which flows through this bottleneck. There are at least three spheres of circulation which meet in the bottleneck : merchandise, measuring instruments, and money. The triad of the three spheres makes it necessary to look at heterogeneous historical sources, not just the publications of the Academies. The requirements of the state, the history of wars, and especially, the history of administration are pertinent. In a sense, the order of the new world of the nation states is crystallised in the bottleneck.

In this paper, I will discuss the importance of these bottlenecks for the fiscal-military nation state and for science with the examples of the two Scandinavian states : Denmark-Norway and Sweden-Finland. The nation states were forced to raise revenue lest they be overrun by their neighbour. Revenue is never easy to come by, as tax payers will not pay unless forced to. The state had limited resources with which to raise taxes, which can be seen from the fact that tax levying was often farmed out. The bottleneck provided an excellent means of coping with limited resources. State officials had merchandise flow through particular bottlenecks, and so the state could deploy its forces just there.

The Scandinavian states are typical. They resemble other Northern European states in that they developed into fiscal-military states with a state administration ticking over even when sovereigns changed. They all established more bottlenecks and expanded the civil service. Southern European states such as Spain and Portugal, or indeed Greece (still a part of the Ottoman Empire) did not, or at least they did so to a much smaller degree. This division between Northern and Southern Europe is also visible in the institutionalisation of science. The bottlenecks are the crucial points for the development of science and a major activity of the national Academies of Science was to find solutions to problems arising at the bottlenecks. The import of the focus on bottlenecks is that the three spheres of circulation mutually impact upon each other and (for the history of science) that the development of science is directly dependent upon the funding by the state and the state's deployment of science in these special locales.

Excise tables of the fees to be paid illustrate the confluence of the three spheres of circulation. The Danish excise regulation for 1779 (Illustration 2 : *Forordning angaaende Consumtionens...*), for instance, lists the fee for barley groats (Byggryn) as 48 shillings per barrel, and for French or Rhinelandish spirits (Brændeviin) as 14 Rixdaler for a hogshead and 4 shillings per " pot " for the domestic variety when sold in Copenhagen (2 in the other towns). Such tables always contain three columns. The first column contains the commodities (grain, spirit, cloth) ; the second column lists the units of measurement (Pund, Lispund, *etc.*) which is different for each commodity, because each commodity is measured with a different instrument ; the third column lists the fee to be paid for one unit of commodity. Taxation tables always present such triads : commodity, unit of measurement, and money.

There is an increase in literature emphasizing the role of quantification and the state. Norton *Wise's The Values of Precision* provides essays giving an insight into many aspects of numbers for the purpose of state administration. Eric Brian's *La Mesure de l'Etat* provides an analysis of the role of numbers in 18th century France. Ted Porter's *Trust in Numbers* discusses the mechanisms of control provided by numbers within administrations, and Rosser Matthews has connected up thumbnail stories about the role of numbers in medicine leading up to the clinical trial. More generally, Bruno Latour's *Science in Action* depicts science generically within terms of circulation of scientists and inscriptions (but not merchandise) and metrology[1]. But noone has yet pointed to the confluence in controlled spaces as the places where almost all

1. M.N. Wise (ed.), *The Values of Precision*, Princeton, Princeton University Press, 1995 ; E. Brian, *La mesure de l'Etat - Administrateurs et géomètres au XVIIIe siècle*, Paris, Albin Michel, 1994 ; T.M. Porter, *Trust in Numbers - The Pursuit of Objectivity in Science and Public Life*, Princeton, Princeton University Press, 1995 ; J. Rosser Matthews, *Quantification and the Quest for Medical Certainty*, Princeton, Princeton University Press, 1995 ; B. Latour, *Science in Action*, Milton Keynes, Open University Press, 1987.

18th century measurement took place. I believe that the concept of bottlenecks could be fruitfully employed in history of science generally.

The paper has three sections : in the first I will describe the context of the military-fiscal state, which will show that the role of indirect taxation (and thus of bottlenecks) increased during the 18th century. It will attempt to put some numbers on the circulation of merchandise. The context of the military-fiscal state will also illustrate the state's motivation to establish such bottlenecks and in particular the deployment of measuring instruments.

The second section deals with these instruments. The main issue is precision, the demand for which increased markedly in the 18th century. This second section will also describe the concomitant development of increasing calibration and the professionalization of measurers (surveyors). Instruments can change with time, they can be fiddled with, or they can simply fail to work properly. Routine calibration enables all parties involved to put more trust in the accuracy of the measurements, as does their use by professional and certified individuals. Two examples of scientific instruments will illustrate the development of precision instruments for the specific problems of measurement at a bottleneck.

The third section deals with money. The history of coinage is not usually seen as a part of the history of science. In fact our disciplinary boundaries normally carve it up into economic history, numismatics, and history of technology. This fact makes coinage a very tricky subject to tackle historically. The literature is diverse and largely incompatible. I wish to argue that the history of coinage is central to the history of science in the service of the state. (For instance, that it was only in the 1660s that the issues of coinage was institutionally separated from weights and measures in Sweden). Coins were produced in mints, in which the highest degree of precision measurement was attained. One could argue that the most sophisticated science in 18th century Scandinavia took place in the mints : under the most closely controlled conditions, carefully executed metal work, chemical assaying and high-precision weighing was performed. Also, in the absence of reliable weights, it was common to substitute them with coins. Money thus literally became a part of the apparatus (balance plus weights) used in the bottlenecks.

In all three sections, a rough indication of the historical development throughout the 18th century will be given ; in all cases showing a steady increase in state machinery, control mechanisms, precision measurement, and stability of value.

THE MILITARY-FISCAL STATE

Until at least 1800, both Scandinavian states were aggressive expansionist military states, constantly interlocked in wars. In the 17th century, Sweden had gained the upper hand, by 1660 controlling most of the Baltic coastal areas. In the 1690s, more than a fifth of Sweden's revenue came from the Baltic provinces, such as Ingria which supplied Sweden proper with large quantities of grain, in 1696 (a year of famine) as much as 800.000 tons[2]. Denmark-Norway having lost most of present-day southern Sweden in 1660 sought every opportunity for revenge. In the Great Northern War (1709-1720), Sweden was faced not only with its traditional Scandinavian rival but also the expanding Russia of Peter the Great and with Poland and Prussia. The Swedish empire eventually crumbled in the face of this massed opposition[3].

The cost of the frequent wars put much pressure on taxation. The period after 1720 increased the fiscal pressures, even in Denmark-Norway which essentially entered into no armed conflict until the Napoleonic Wars, because it was almost always on the brink of war and arming heavily. Sweden-Finland was engaged in constant warfare in a futile attempt to contain Russia. Russia's emergence as a Baltic power rendered control of Baltic trade increasingly difficult for the Scandinavian states and their grip on bottlenecks there was broken[4]. Accordingly, both Scandinavian states had to turn their efforts towards increasing revenue and economic strength from within. They became capable of levying ever higher taxes through creation of domestic bottlenecks, refinement of the state machinery, helped also by a buoyant economy. The price of grain rose considerably towards the mid-18th century, and towards the end of the century, modernization of agriculture (enclosure and privatization of land followed by more intensive farming techniques) was beginning to produce higher yields too. This boom led to an emergence of a private sector of more than negligible proportion. The availability of credit increased.

In 1660, the only accumulation of capital had been in the hands of the king, and to a lesser extent in the church coffers. The landed nobility had received dues in kind and not in money. Land was not bought or sold to any significant degree : peasants tilled the land owned by the gentry against an annual manorial due. By the mid-18th century, banks were coming into existence, and exchange involving money was becoming increasingly common. Hitherto common and open land was being parcelled out, enclosed and owned by the farmer tilling it. This momentous development relied very much on the land becoming a commodity for which the evaluation and pricing of land was a pre-

---

2. D. Kirby, *Northern Europe in the Early Modern Period - The Baltic World 1492-1772*, London, New York, Longman, 1990, 257.

3. H. Becker-Christensen, *Dansk toldhistorie II- Protektionisme og reformer - 1660-1814*, København, Toldhistorisk Selskab, 1988 ; Knud J.V. Jespersen, *Danmarks historie - Bind 3 : Tiden 1648-1730*, Copenhagen, Gyldendal, 1989 ; S.A. Nilsson, *De stora krigens tid - Om Sverige som militärstat och bondesamhälle ; The Era of the Great Wars - Sweden as a Military State and its Agrarian Society*, Uppsala, Almqvist & Wiksell International, 1990.

4. H. Becker-Christensen, *Dansk toldhistorie II* ; Jespersen, *Danmarks historie - Bind 3* ; E.F. Heckscher, *An Economic History of Sweden*, Cambridge, Mass., Harvard University Press, 1954 ; G. Behre, L.-O. Larsson, E. Österberg, *Sveriges historia 1521-1809 - Stormaktsdröm och småstatsrealiteter*, Stockholm, Esselte Studium AB, 1985.

requisite. Both the increasing privatisation of land of the 18ᵗʰ century and the increasing role of indirect taxation required a wider network of state control with surveying and other measurement.

John Brewer has shown how throughout the 18ᵗʰ century, indirect taxation became the main form of taxation in Britain[5]. The excise, in particular, became an organisation of extraordinary efficiency. Essential to the funding of the many wars of the period. He emphasises as a British specialty : the lack of venality but also the particular form of accountancy which developed and which made it possible for the Treasury to see with clarity what the total income and expenditure of the state was. This allows the historian to follow the British case with some ease. While the Scandinavian revenue figures are less informative than the British ones, it can at least be said that the financing of wars was central to the fiscal systems, that the state tentacles reached further and further into the nooks and crannies of society for the purpose of levying indirect taxes, that this provoked protest, and that the state promulgated standards in order to quell protests and have the operation proceed smoothly[6]. For the importance of the indirect tax, the following points can be made. The most serious public sector borrowing requirements fell within the years of the Prussian war[7]. Excluding loans, it amounted to appr. 20% of the state's revenue, and to about 70% of the land tax. Indirect tax constituted a fair slice of the state's revenue (but not nearly as much as in Britain). Also, it proved a more flexible form of income than the land tax. From the peace time of 1722 to the war time of 1758, the land tax rose by 18%, whereas the indirect tax rose by 54%. The same fiscal pressures apply to the Danish-Norwegian state ; its debt forced ever higher by the mere threat of warfare[8] : and indirect taxation provided growing revenue throughout the 18ᵗʰ century, just as was the case in Britain[9], the importance of bottlenecks is borne out by the statistics. Town gates provided much better control than turnpikes which were more easily circumvented. In the early 1670s, characteristically, Danish provincial towns levied appr. 88.000 rigsdaler annually. By comparison, the number for Copenhagen is 69.000, and for the countryside it was only 48.500.

The excise duties of both Scandinavian states reveal the sophistication of the systems of measurement. The Swedish Purchase and Consumption Tax Bill of 1756 distinguished between five different forms of taxation :

5. J. Brewer, *Sinews of Power - War, Money and the English State 1688-1783*, London, Routledge, 1989.

6. K. Åmark, *Sveriges Statsfinanser 1719-1809*, Stockholm, P.A. Norstedt & Söners Förlag, 1961, 400-414.

7. *Ibid*.

8. Figures taken from O. Feldbæk, *Danmarks økonomiske historie 1500-1840*, Herning, Systime, 1993, 152.

9. For Britain, *cf.* Brewer, *Sinews of Power* ; For Denmark-Norway, *cf* : Becker-Christensen, *Dansk toldhistorie II*.

1) The domestic customs duty was to be levied when agricultural or manufactured goods were brought to town, market or to certain designated adminis-
trational areas (including the mining districts). The tariff was paid per barrel :
   - wheat flour : 8 öre
   - wheat, rye flour, and peas : 5.5 öre
   - rye, barley, malt : 4 öre
   - oats : 2 öre
2) The port excise duty.
3) Home excise, levied on home brewing and home baking. The fee had to
be paid at the local excise office. For brewing it was measured by the quantity
of malt bought.
4) A sales excise, to be paid by brewers, bakers, butchers, meat traders, *etc.*
5) Mill duty[10].

The duties were set nationally and it is to be noticed again that specifica-
tions were always in terms of the trinity : commodity, a unit of measurement
and a price. The tax payer brought the commodity to the bottleneck, a location
where state-disseminated weights and measures were found. Here a measurement was performed by a civil servant, to some extent trained and certified by
the state ; and the tax was to be paid in coins also disseminated by the state,
the value of which was vouchsafed for by the state.

When the state promulgated standard weights and measures, it was explicitly in order to control abuse. The state defined ground rules for the many professions which dealt in day-to-day measurement : measurers, weighers,
carriers, port controllers and firewood measurers[11]. It repeatedly decreed that
standard metropolitan weights and measures be used exclusively in customs
affairs. In 1687 it was decreed that all local weights and measures be destroyed
and replaced with standards[12]. The repetition of such legal measures indicate
that the position on the ground continued to be under less than full control.
Bottlenecks ameliorated this problem.

MEASURING INSTRUMENTS AND THEIR USE

Accounts kept in the Copenhagen Town Hall enable an estimate of the calibrations carried through and secondary standards sold, thus giving an indication of the spread of calibrated measurement in Denmark-Norway of the time.
The magistrate's Chamber of Calibration received just over 1.637 rigsdaler in
6 months of 1772 for sold weights and measures and calibration fees. Calculating from the prices and fees given this would correspond to, for example,

10. K. Åmark, *Spannmålshandel och spannmålspolitikk i Sverige 1719-1830 - Akademisk
avhandling*, Stockholm, Isaac Marcus' Boktryckeri-Aktiebolag, 1915, 81-88.
   11. *Cf. Anordning om Maalere, Veyere, Vragere, Haufne-Fogder oc Faufnsettere*, Copenhagen, Bockenhoffer, 1683.
   12. H. Becker-Christensen, *Dansk toldhistorie II*, 126.

the sale of 96 fitted salt barrels, 96 tin pots, 96 sets of 10 *lispund* weights, 96 sets of 1 *lispund* weights, plus the calibration of 96 salt barrels, 96 oil barrels, 96 tin pots, 96 tin jugs, 96 iron rods (1 *alen* long), and a 10 *lispund* metal weight. And this activity increased : in 1794, the income for calibration and sale of calibrated weights and measures had increased to just over 2.795 rigs-daler[13]. These numbers give an indication of the numbers for the late 18th century. The Copenhagen Town Hall has no similar records, allowing the historian to compare with earlier times. But it is possible to sketch the historical development by utilizing the rules and regulations related to weights and measures. This section will give an outline of these along with an account of the professionalization of measurers and then turn to two specific instruments as examples of the concerns for precision measurement in the context of taxation.

Danish weights and measures were reformed in 1683, when a regulation was devised in consultation with Ole Rømer[14]. Rømer had intended all weights and measures to be reducible from one standard length. Standard weights, for instance were given by cubic vessels of standard length, containing pure water. But the reality of weights and measures remained a far cry from Rømer's grand plans for order[15]. The 1683 regulation prescribed national standards and was followed up by supporting regulations intended to ease through the introduction of the new standards on a national scale. In order to ensure that weights and measures were indeed standardised, the Copenhagen magistrate was given monopoly of their sale. But craftsmen resisted this monopoly vigorously also after Rømer succeeded in wresting the monopoly from the magistrate in the late 1680s. When it was clear in 1687, that old measures were still being used, their destruction was prescribed but to no avail. Eventually, the king had to interfere in a dispute which dragged on until 1698 between the pewterers, the magistrate and Rømer. A new regulation now gave the Copenhagen magistrate and 4 provincial towns monopoly on sale of standards[16]. The difference between legal prescription and actual practice remained, but it was reduced. From 1698 onwards the Copenhagen magistrate began to take advantage of its monopoly. Accounts reveal that the magistrate made a nice profit over the cost of labour for cooper and smith when selling weights and measures. Citizens could also bring their weights and measures to the Town Hall for calibration against a fee[17].

13. Københavns magistrat, MC600 : 1696-1795 : Justervæsen, Copenhagen Town Hall.

14. " Ole Rømer ", *Dictionary of Scientific Biography*.

15. K. Meyer, " Dansk Maal og Vægt fra Ole Rømers Tid til Meterloven ", in *Beretning fra Meterudvalget om dets Virksomhed i Tiden fra dets Nedsættelse den 9. Juli 1907 indtil den 31. Marts 1914*, Copenhagen, J.H. Schultz, 1915, 57-91, esp. 57-70. *Cf.* also A. Hægstad, *Mål og vægt i Danmark 1283-1983 : den legale metrologi gennem 700 år*, Copenhagen, 1983.

16. *Ibid.*, 70-71.

17. Copenhagen Town Hall, Rådstuearkivet, Justerkamrets dokumenter 1696-1819. Also : K. Meyer, " Dansk Maal og Vægt ", 74-76.

By the 1740s, the task of Swedish calibrators was to some extent coordinated with local prefects. In the 1750s, just in time for the expensive war with
Prussia, the enforcement of the laws was aided by the specification of punishments. All instruments (barrels, mugs, rods, balances, *etc*) were to be calibrated and hallmarked. Any trader possessing instruments not yet hallmarked
was to have them confiscated and fined. Anyone using the wrong measures was
to be put in the stocks for an hour, and to receive ninety beatings. Calibrators
faking a hallmark were sentenced to death. Anyone changing a hallmarked
instrument was fined and " dishonoured ". In the same period the problem of
wear and tear was regulated. Weights in particular were known to change with
use, and the customs asked Jacob Faggot, the Head of the Board of Surveyors,
for advice. As a result the Board came up with guidelines for corrections to
wear and tear. Also, topping up was prohibited and exact recompense was prescribed in some cases. In the instruction of 1783, the right of anyone to
become a calibrator provided he (gender used advisedly) passed an exam,
came to an end. Now, calibrators had to be appointed by the local administration. In the 1790s, the tasks were routinised. For instance, in Stockholm, every
month belonged to the calibration of instruments employed in particular trades.
In January all butchers, fish mongers and iron mongers were to have their
instruments calibrated. In February it was the turn of brewers, bakers, distillers,
and vinegar makers, and so on[18].

As Witold Kula has shown with abundant material from 18th century Poland
and France, it is easy to get different results even when measuring with the
same instrument. The local diversity in measuring practice which surveyors
faced was great[19]. Not only were weights and measures different in different
localities, but the practice of topping up after weighing or measuring was also
common. The practice of exchange was not just a question of determining the
equivalent value of two wares, or that of a ware in money, but an almost ritu-
alised encounter. In Sweden, topping up could sometimes add 20% or more to
the measured volume. Kula has shown that within local communities everyone
knew the rules of the game, but with increasing circulation of traders and merchandise, the local varieties became increasingly problematic. The state, which
was attempting to coordinate the activities across many local communities had
an incentive to standardise, but locals resisted the new ways.

The human measurer thus introduced a moment of uncertainty into the aim
of stabilizing value. The solution to this problem proffered in the 18th century
was similar to the solution for instruments and coins. Education of measurers
amounted to the routinization of standard procedures, and professionalization
amounted to certification in analogy with the seal of the sovereign.

18. *Svenska Lantmäteriet 1628-1928*, 2 vols, Stockholm, Sälskapet för utgivandet av lantmäte-
riets historia, 1928, especially the articles by J. Svärdson, H. Juhlin Dannefelt and E. Williams.
    19. W. Kula, *Measures and Men*, Princeton, New Jersey, Princeton University Press, 1986.

Looking back from the mid-18th century, Jacob Faggot, the head of the Board of Surveyors, argued that before 1680, surveying had been performed more as a craft than a science, and that it ought to be done by " washed hands ", referring to the class differences in cleanliness, and the division between intellectual and manual labour.

What follows is a description of the *de facto* and *de jure* professionalisation between 1660 and 1800 — in terms of the creation of a monopoly, of training, and control. The historical sources allows this to be illustrated particularly well for Sweden-Finland, which thus takes pride of place. In mid-16th century Sweden, there were no professional surveyors. Bailiffs, the county chief and a committee of local peasants carried out land evaluation. Local men or boys were employed measure with rods. Complaints from peasants about unjust taxation persisted throughout the century. A Board of Surveyors was set up in 1628 specifically in response to the complaints, and with the aim to make taxation more just and thereby facilitating its collection. The new surveying was to be coordinated by the prefects. All measurement protocols were to be submitted to the *kammarkollegium* Area and quality of the soil (measured by yield of number of haystacks) was to be reported along with the owner's name and number of grown-ups living on the land. A land register was to be collated from this. Initially, maps were not drawn. In the last decades of the 17th century, when the *reduktion* (the state confiscation of nobility-owned land and its subsequent parcelling out to soldiers) was at its peak, surveys were carried out all over Sweden. New regulations were introduced : in 1687, the king prescribed that surveyors take an oath as qualification for this kind of work. It was also decreed that houses and properties be searched as part of the evaluation, and that an inventory be established. Surveys, and by extension evaluations, were controversial, so that four additional local civil servants were to participate. A report of measurements, evaluation, and drawings was to be submitted to the *kammerkollegium* for each unit of land surveyed.

We have repeatedly seen that several trustworthy individuals apart from the surveyor had to take part in the measuring process for the result to be trusted. Who measured mattered and the regulations were getting increasingly explicit about whom to involve. In addition, the question of measurement technique also gained in importance. Some of the early instruments were very simple. Mostly, ropes were employed. It is only towards the end of the 17th century, when the *reduktion* was under way, that geometrical instruments were used. Geometrical was the term employed to distinguish the instruments measuring angles from rods and chains measuring lenghts. These instruments involved more complicated geometrical calculations.

In the 1660s, weights were for the first time separated from the issue of coinage. An inspector was appointed by the *kammarkollegium* to oversee all large balances and weights of the realm, except for the small ones intended for gold and silver which continued to be under the auspices of the minting examiner (*riks-guardien*). In 1665, a declaration was posted that only certain balances and weights were to be employed, but no additional hands were employed to test existing instruments. Calibration was to be the responsibility of the general administration, which was naturally busy with other matters.

Around 1700 a number of initiatives were taken. Surveyors were to be examined and licensed. All rods, chains, and ropes were to be calibrated against the Stockholm unit of length. Also, the intellectually more demanding instruments to be employed by " washed hands " such as astrolabes, quadrants, compasses, spirit levels, proportional circles (invented by Galileo), rulers and similar instrument were to be introduced.

In the early part of the 18th century peasant grievances about the abuse of weights and measures increased. In response, the parliament debated the pros and cons of a corps of well-qualified calibrators (*justerare*) spread out over the entire realm. In 1735 the government decided to add weights and measures to the responsibilities of the Board of Surveyors. Market towns were to have a calibrator taken under oath, who also had to pass an exam at the Board of Surveyors. Calibrators were to be remunerated through fees, and it was to become compulsory for traders to have their instruments calibrated. As a result of Scanian peasant resistance in the 1750s and 1760s, a new instruction for surveyors was introduced in 1766. It prescribed in much greater detail just what was to be measured and reported back to the central administration. In addition, the peasants involved had the right to appoint independent " reliable " (what this implies is not specified) observers when the quality of the soil was to be evaluated, except when the evaluation was simultaneously done for the purposes of determining the tax level. These independent observers were to try to mediate and resolve any dispute, before the case was referred to local courts. Because of these practical organisational problems, the instruction of 1766 reiterated the need for the introduction of calibrated instruments. Now, the practice of surveying was prescribed in even greater detail. There were rules for the setting up of corner stones, and rules for dealing with the topography of the land. Reconnoitering in advance of the actual measurements was emphasised. And there were further rules for the way in which maps were to be drawn and information to be recorded. In the winter and in all hours of leisure, the surveyor was to make fair copies of all his drafts. The kind of paper, pen and colour to be employed was prescribed too. By 1783, a new surveyors' instruction specified further the process of evaluation of land quality. Fields, meadows, newly cultivated land, forest etc. were now to be evaluated separately. Within each kind there were several gradations for the possible value. At the same time the ways in which compensation for exchange of superior against inferior land could be paid (manure, labour, or money) was determined.

In the rest of this section, I will examine two sample instruments which both illustrate the concerns of precision measurement within a network for the purposes of taxation. The first, described in a publication in the *Transactions of the Copenhagen Academy* illustrates the production of a metropolitan standard aiming to improve the levying of an alcohol excise[20]. It addresses the problems of precision and reliability and calibration in the metropolitan setting in order that the instrument be used elsewhere. It was entitled " On the means of examining, testing and evaluating all spirits traded, both in terms of their measure and their quality, *etc* "[21]. The author, Franz Heinrich Müller, stated that the Customs Office had set him the task of making an instrument for the purpose of assessing alcohol content, in order to " measure their value in money ", while also emphasising that every trader can use it, since he must know the value of merchandise to avoid harming either himself or his customer and retaining the faith of the public. As we shall see, the parallel with the second instrument examined below, Ekström's grain tester, is extensive : precision measurement employed for the purpose of evaluation ; the making of a standard instrument initiated and paid for by the state but intended also for the market.

Müller refers to the measuring rods of standard size usually employed to measure the content of barrel by holding them diagonally within the barrel. The much more taxing problem was the alcohol content. His solution was a silver rod which was to be dipped into the liquid in question (see illustration 3 : Scala for en Brændevins Pröver). The portion submerged indicated the alcohol content, and the silver rod was to be given markings denoting this alcohol content. These markings were applied in a metropolitan laboratory where the purest water and alcohol available were mixed in varying proportions and the dip of the silver rod tested. The rod is depicted on the left of the illustration and the columns on the right refer to I : the total quantity of liquid ; II : the mix of water and spirit. The original calibration of the rod was performed at 0 degrees Reaumur (already fairly advanced, requiring control of temperature and a thermometer calibrated on the Reaumur scale). The calibrations on the rod with the numbers in column III, referring to the quotient of water and alcohol. However, this quotient was different for moderate and hot weather (given in columns IV and V, respectively). These quotients were also tested at the time of the original calibration.

Müller emphasised that other methods of alcohol content were less useful : distillation was too expensive ; burning required absence of wind ; a hydrostatic balance was fragile, slow and required to be used indoors ; judging by

20. The alcohol excise was introduced in 1688, *cf.* H. Becker-Christensen, *Dansk toldhistorie II*, 256 and it became increasingly important for the state coffers throughout the 18th century.

21. F.H. Müller, " Om Maaden og Midlerne at undersøge, prøve og vurdere alle i Handelen forekommende Brændevine, i Hensigt til deres Beskaffenhed, som Maal, med videre ", *Skrifter som udi det Kiøbenhavnske Selskab af Lærdoms og Videnskabers Elskere ere fremlagte og oplæste*, New Series, 3 (1788), 202-219 ; he completely changed the design later : " Nøiere Oplysning og Forbedring vedkommende Brændeviinsprøveren og sammes Anvendelse ", *Skrifter som udi det Kiøbenhavnske Selskab af Lærdoms og Videnskabers Elskere ere fremlagte og oplæste*, New Series, 5 (1799), 71-81.

the bubbles created by shaking was unreliable. His rod thus " travelled " better than any of the rival candidates. Still, he did emphasise that the rod was subtle and must be protected from knocks and falls with a padded container. The rod could be used to detect fraud almost instantly. But not all fraudulent techniques were detected by it. One such technique was to heat the liquid just before the measurement was to take place, and thus temperature must be carefully checked. The trust inspired in the value of the merchandise by the spirit tester was to provide savings in transport, insurance and other costs.

This silver rod spirit tester thus illustrates very clearly the issue of calibrating an instrument which can be used with ease out in the field where measurement and evaluation takes place. It also shows the problems of control, authority and fraud.

Towards the end of the 18th century, many such instruments were in demand. The demand for a local precision instrument maker thus became acute. The surveys required good sextants, mural quadrants, geographical circles, levelling instruments, compasses, chains of measurement, good quality paper, etc[22]. Customs and excise required a variety of instruments such as Müller's silver rod. But no instrument maker capable of producing the kind of quality which was the norm in European centres lived in Copenhagen, nor were the requisite skills to be found with the already present crafts. It required that someone gathered the experience abroad and returned to set up shop in Copenhagen[23].

The second instrument, a grain tester devised by Daniel Ekström, is equally informative about approaches to measurement (illustration 4 : " Tab VI "). Its purpose was to regularize the measurement of the quality of grain seed by comparing its weight and volume. It was not just designed for official use ; Ekström emphasizes that any tinsmith could make it, and he recommended its widespread adoption. It would improve accuracy of testing even when reproduced with only moderate tolerances, because it required no close measurement of volume ; in conjunction with an accurate balance and weight, it allowed reliable comparisons between grain samples.

The way in which the grain is poured is critical since it affects packing ; in mill practice it was considered cheating to pour grain into a barrel from higher than the hip. Similarly, when testing samples of grain, seed would be dropped carefully into the container from cupped hands as shown. Again, the way of pouring the seed could be employed to pack the seed differently. Ekström's instrument is intended to eliminate the uncertainty of the use of hands in the normal measuring process (fig. 1 on the left hand side of the illustration is

22. K. Nielsen, *Hvordan Danmarkskortet kom til at ligne Danmark - Videnskabernes Selskabs opmåling 1762-1820*, Århus, 1982, 12.

23. D.Ch. Christensen, " Spying on Scientific Instruments - the Career of Jesper Bidstrup ", *Annals of Science*, 52 (1994).

Ekström's instrument, the hands on the right simply indicate the usual procedure which he aims to eliminate). The container *ab* has an adjustable floor *c*.
The container *ab* is now put underneath another container *ghik*, which has a
hopper at the bottom that can be opened and closed. With the hole closed, the
seed is poured into *ghik*. The rod *lm* determines the height between *ghik* and
ab and can be set to a standard distance. Thus, to whatever capacity the container is set, the distance between hopper and floor remains the same. By open-
ing the hopper, the seed then falls into *ab* in a standard way unlike the pouring
from cupped hands.

In preparation for the measurement, the seed was to be dried to prevent the
water content from influencing the measurement. Ekström suggests standard
ways and lengths of time for the drying process. Once the grain samples to be
compared are equally dry, a known weight of the first grain type is poured into
the container until the grain fills the container to overflowing, and the surplus
is levelled off, flush with the rim, using a specially constructed tool which
Ekström describes in some detail. The container is then emptied, without
changing its capacity ; grain of the second type is then also poured in, until it
also fills the container exactly to the rim. It is interesting that Ekström is not
satisfied to use a container of fixed capacity. Probably, this is so that the container can be adjusted to the standard weights for each kind of grain.

With many grain testers of this kind, the quality of seed could be ascertained
everywhere. Ekström referred to the need for distributing grain testers by mentioning the most recent Royal decree on barrel measures. The state actively
supported this kind of work ; in fact Ekström performed the experiments on the
order of the *kammarkollegium* at the royal grain depot. Ekström also claimed
that the instrument would help to mediate in international grain trade, referring
to Dutch standards and the requirements of the Dutch market[24].

MONEY

Coinage was always crucial to the collection of taxes, as far back as Roman
times[25]. The value of coins was forever in dispute, and there were repeated
attempts to standardise them. Furthermore, the precision of the value of coins
continued to be an issue and the state mostly took great pains to keep coins as
uniform and standardised as possible. The process of minting was highly specialised, involving both precision weighing, chemical expertise, and extreme
control. The value of the coin produced under these circumstances was vouchsafed for by the image of the sovereign on the coins themselves, very much

24. D. Ekström, " Beskrifning På en Spanmåls-profvare, inrättad efter Svänskt Mål och Vigt. ",
*Kongl. Svenska VetenskapsAcademiens Handlingar för Månaderna Julius, Augustus, September,
År 1753*, 224-241.

25. J. Porteous, *Coins in history - A survey of coinage from the Reform of Diocletian to the
Latin Monetary Union*, London, Weidenfeld and Nicolson, 1969, 15.

like the standard weights, measures and rods. There was also a geography of measurement with a hierarchy of precision, the highest precision achieved at the mints, the lowest in a provincial marketplace. Measurement and evaluation is thus intimately involved in coinage and many of the problems are the same as in weights and measures. Bottlenecks were also special points where counterfeit coins could be detected.

The main techniques for testing gold and silver content in coins were widely disseminated all over Europe even in the Middle Ages[26]. Weighing was one way of assessing a coin's value, but sub-standard coins could be rendered heavy by an admixture of lead. A more sophisticated measure consisted of the hydrostatic weighing, that is weighing consecutively in air and water which, as we would now say, detects lead by determining the specific weight of the amalgam[27]. The sophisticated test which was known to goldsmiths all over Europe in the Middle Ages and remained an integral part of minting practice throughout the period concerned here was a full-blown chemical analysis. Towards the end of the 18th century, the trade secrets of coin assaying were published in book form[28]. The sample was placed in a small container made of clay. Aqua regia was added and the container was heated to a high temperature in an oven. The impurities were absorbed into the walls of the container leaving the noble metals to be weighed.

These analytical and measuring practices spread throughout society wherever money changed hands. Simple, cheap balances were more widely employed, and many traders would employ hydrostatic balances (especially towards the end of the 18th century), whereas the costly and time consuming full-blown chemical analysis was restricted to only a few places, such as mints. The state attempted to counter coin clipping and the filing off of metal dust with harsh punishments but also by setting limits for the acceptability of coins. They were to possess weights within particular limits. If found too light, they could not be accepted as tax payment. Similarly, the content of precious metal in the coins had to be within certain limits. This was called the *remedium*[29].

26. Sir J. Craig, *The Mint - A History of the London Mint from A.D. 287 to 1948*, Cambridge, Cambridge University Press, 1953. For the so-called trial of the pyx, see esp. 394-397.

27. Ekström also had such instruments in his repertoir ; *cf.* A memorial by J. Faggot to the Swedish estates : " 6.2 Memorial til Riksens Ständers manufactur och Handels deput. ang. mathematiska Instrumentmakeriet, 29 martii 1739 ", Lantmäteristyrelsens expeditionsböcker B1 :14 1731-1740, Lantmäteriverkets arkiv, Gävle ; *cf.* also M.A. Crawforth, *Weighing coins - English folding balances of the 18th and 19th centuries*, London, Cape Horn Trading Coy., 1979, who gives a good impression of the increasing distribution of precision balances towards the end of the 18th century. M. Daumas also covers precision balances on pp. 221-227 of his *Scientific Instruments of the Seventeenth and Eighteenth Centuries and Their Makers*, London, Batsford, 1972.

28. S. Haase, *Eröffnetes Geheimnus der praktischen Münzwissenschaft*, 1762 ; S. Haase, *Vollständiger Münzmeister und Münzwardein*, 1765 ; J.O. Ruperti, *Das Probiren, in so weit diese Wissenschaft zu dem Münzwesen nothwendig gehöret*, Braunschweig, 1765. Note that Ruperti's title refers to assaying as a science.

29. e.g. *ibid.*, 179.

The precision with which the mint produced coins within the limits was extremely important. If the content of precious metal was too high, the coins would disappear from circulation and be melted down ; if it was too low, it would not inspire trust (especially as the state itself might not accept it is tax payment) and it would fall below its stipulated value causing inflation. If coins were produced with too large a spread in the precious metal content, the more valuable ones would disappear. This was called " seigern " in German and the process was denoted " Gresham's law " in English[30]. The Crown thus had a great interest in the production of coinage with very precise contents and weights and in preventing tampering with coins, both for the purposes of having money with which tax could be paid, and for the purposes of enabling trade and avoiding inflation[31].

The value of coins was vouchsafed for by the organisation of the mint in a similar fashion to the organisation of a laboratory : restriction of access, use of sophisticated scientific instruments and accounting practices, enabling the sovereign to keep some degree of control over work in the mints without actually having to be present at the time of minting. Indeed, the mint is also a bottleneck. Furthermore, the work of the Master of the Mint was surveyed by a Guardian, employed for that purpose. The regulations and control techniques at the Mint got ever more sophisticated in the 17th and 18th centuries. At the Royal Danish Kongsberg Mint, the instructions for the Master of the Mint and the Guardian in 1628 and 1629 were very brief by the standards of a century later (the text is roughly a fourth in length). The Guardian of 1629 had mainly to be observant and diligently assay the amalgam before it is struck[32]. By contrast, the instructions for the Guardian in 1730 consisted of 14 paragraphs. He was now obliged to take samples several times during the minting process and it was prescribed who was to be present. He was then to analyse the metal contents in his own laboratory. The locking up and sealing of samples, the accountancy and report required were much more detailed than a century earlier. He was now also allowed to calibrate samples brought to him against a fee, while remaining under his oath of fidelity to the king, thus very much resembling the role of the Stockholm and Copenhagen magistrates[33]. Again, trust, authority, and detailed regulation combine to vouchsafe for a stable value.

The organisation of the Swedish Mint was similar. A manuscript which has been dated to 1720, refers to the practice of minting, the control functions of

30. A. Luschin von Ebengreuth, *Allgemeine Münzkunde und Geldgeschichte des Mittelalters und der neueren Zeit*, 2. ed., München & Berlin, R. Oldenbourg, 1926 ; Craig (note 27).

31. This did not prevent sovereigns from attempting to make a profit from coinage, by issueing substandard coins, but state and trade always paid for this through the obstacles put in the way of smooth collection of tax and exchange of merchandise. *Cf.* Craig, *The Mint, passim.*

32. B.R. Rønning, *Den Kongelige Mynt 1628-1686-1806*, Norges Bank/J.W. Cappelens Forlag a.s., no place, 1986, 298-299.

33. *Ibid.*, 310-314.

the Guardian, and the bookkeeping procedures in virtually the same terms as the Danish one of 1730[34]. In it, the authority of the sovereign is addressed : " Those who counterfeit should receive harsh punishment, not just for reaping a large profit but especially for profaning so shamefully the royal prerogative... Noone should disrespect the royal mint ; rather one ought to respect the country's honour and the royal prerogative, not tread it under foot. The ancient Romans considered throwing a coin with the sovereign's image into the latrine a *crimen læsæ Majestatis* "[35].

The social organisation of the mints and their use of scientific instrumentation was a result of this concern of the state. To glimpse the kind of precision work performed in a mint, we can inspect an 1806 inventory of the Kongsberg mint. From it we can see that precision measurement took place here, along with chemical analyses of various kinds. One room was called " laboratorium ", where a *Probeer-Vægt* (assaying balance) and two *Proberovne* (chemical ovens) were found. Next door was an iron balance with copper bowls and weights of various sizes. In another building was a room with two rough, five precision, and two very high precision balances. Many rooms contained ovens and various kinds of tools for handling materials in the ovens. There was also a *Kontoir* (comptoir or office) where containers with locks were kept. Many doors had locks bearing witness that this was a place with highly restricted access just like modern laboratories[36].

The publications of the Academies also dealt with precision work in minting, as mentioned above. For example, Müller (the same person who devised a spirit testing instrument) depicted many tricks of the trade, analysing various aspects of minting, such as the dimensions of the oven, the best means of producing vials and containers, the impact of various ways of stacking the coal[37]. The analysis was explicitly intended to improve precision.

CONCLUSION

This account shows that the routine measurement in 18th century Scandinavia took place at bottlenecks provided by the state for fiscal purposes. The bot-

34. The manuscript is entitled : " Humble report on the Mint " (Underdån ödmiuk Relation om Myntet), and is reprinted in K.-A. Wallroth, " Sveriges Mynt 1449-1917, bidrag till en svensk mynthistoria meddelade i myntdirektörens underdåniga ämbetsberättelser ", *Numismatiska meddelanden utgivna av svenska numismatiska föreningen*, 12 (1918), 177-201.

35. *Ibid.*, 178.

36. *Inventarium til den kongelige Mynt paa Kongsberg*, dated 11. januar 1806, Rigsarkivet (State archives), Copenhagen, Finanskollegiet, journalsager, jnr. 503, 1806 ; also printed in Rønning, *Den Kongelige Mynt*, 323-329.

37. F.H. Müller, " Sølvets Prøvelse til Nytte for Mynte-Væsenet og Sølv-Handelen ", *Skrifter som udi det Kiøbenhavnske Selskab af Lærdoms og Videnskabers Elskere ere fremlagte og oplæste*, New Series, 2 (1783), 153-173 ; and " Om Guldprøvens nøiagtigste Omgangsmaade til Nytte for Myntvæsenet og Guldhandelen ", *Skrifter som udi det Kiøbenhavnske Selskab af Lærdoms og Videnskabers Elskere ere fremlagte og oplæste*, New Series, 4 (1793), 1-28.

tleneck provides the focus in which the three spheres of circulation (merchandise, measuring instruments, and money) are concatenated. The bottleneck is thus the prime place where the practice of measurement is most commonly found in the 18th century. In Southern Europe, taxes continued to be levied primarily on land, and the nation states did not develop administrations and civil services with the same speed. There was no development of Excises with their routine evaluation of merchandise, and the Academies of Science (Greece did not have one) also languished by comparison.

18th century (and maybe also later) history of science should be seen in the context of the state. The state was involved in most of what has been discussed here. The personal links are explicit. For example, Jacob Faggot, the main force behind enclosure in Sweden, was simultaneously Head of the Board of Surveyors (a part of *kammarkollegiet*, the fiscal department of the government), promoter of the instrument maker, Daniel Ekström, and Member of the Academy of Science. The state was involved in the building of town walls, town gates and other bottlenecks. The state had soldiers posted in such places. The state set up calibration networks and had a hand in the professionalisation of surveyors. Calibration was accompanied by certification : the stamp of authority (of the sovereign of the state) was embossed on coins and instruments. The state employed instrument makers (Ekström was purveyor to the Board of Surveyors) and paid for their education. The state paid the Academy to provide solutions to problems arising at the bottlenecks (in the case of Sweden, the state guaranteed that the Stockholm Academy had a monopoly on almanacs within the reign, which turned out to be quite a money-spinner).

The concatenation of the three spheres of circulation redefines all three. In order for smoothness of operation to take place in the bottlenecks, value must not be an issue leading to conflict. Calibration stabilizes value. The procedure at mints stabilizes value. The stamps of authority stablize value. The more people trust the instruments, the coins, the categories of merchandise, the more stable evaluation in the bottlenecks will be. That coins and instruments are shaped by the threefold concatenation should be obvious. The reconfiguration of the merchandise is akin to the process of commodification described by Marx and splendidly illustrated by Schivelbusch[38]. Marx defines the commodity as the uprooted piece of merchandise which is devoid of any social meaning except as something to be exchanged. As we have seen, the measurement in the bottlenecks provides a stable evaluation of merchandise and configures it

---

38. " This locational movement — the bringing of the product to the market, which is a necessary condition of its circulation... could more precisely be regarded as the transformation of the product *into a commodity* " (K. Marx, *Grundrisse, Foundations of the Critique of Political Economy* (London, 1973), 534, italics in the original). Quoted in W. Schivelbusch, *The Railway Journey - The Industrialization of Time and Space in the 19th century*, Berg, Hamburg, New York, Leamington Spa, 1977, 1986, 40. Schivelbusch brilliantly unearths the diverse cultural impact of increased circulation as a result of the railways.

in particular categories. Marx thought of the commodity as defined by exchange in the market. This paper suggests that one could think of the commodity as defined by the state.

In all cases of evaluation, there were different levels of exactitude. Precision was geographically differentiated. In the case of coins there was weighing, hydrostatic weighing, and chemical analysis. In the case of surveying, there were rods or chains, geometrical analysis on the basis of simple determinations of angle, and high precision surveying instruments. In the case of merchandise, there were simple rods or balances used with coins, there were copies of copies of standard weights and measures, there were standards kept in the capital and finally there were high precision scientific instruments. The level of exactitude was related to the locale. After all, high precision was not always required. It was expensive, demanded great care and controlled surroundings. Low precision was cheap, quick and possible almost everywhere. The highest precision was to be found in the capital where standards were generated under great control. Copies of the standards were promulgated and used in the provinces under diminished control. The bottleneck provides the means for approximating the conditions to those achieved at a standards setting location in the metropolitan centre.

The bottleneck could be seen as a pre-history of the laboratory in that they are places where control of the environment enables precision measurement. Labs are now more specialised and they do not always evaluate commodities. That is one more reason that their economic role is not always very explicit. But it would be possible to examine the economic role of a great many modern laboratories in which evaluation is performed as a part of state regulation. In these cases, the economic role is similar to that of the 18th century. And standards (such as weights and measures) certainly continue to be of great importance in the modern economy.

ILLUSTRATION 1

Vestenport med vagten i begyndelsen af 1800erne
Posten lå på den nuværende Rådhusplads

ILLUSTRATION 2

ILLUSTRATION 3

Tom III pag 20

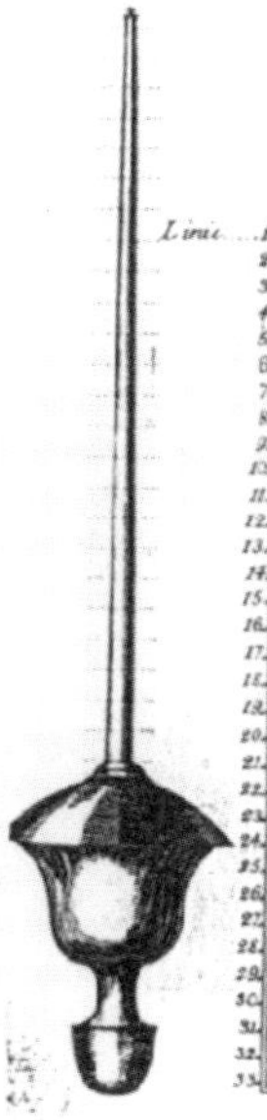

Scala
For en Brændevins Pröver
indrettet til Brug for
Vinteren, Sommeren og tempereret Veierligt

Linie	I Quantitetet	II Vand	Spiritus	III Kold	IV Tempereret	V Hed
1	[illegible]	6	36	36	41	46
2	41	6	35	35	40	44 [illegible]
3	40	6	34	34	39	43 [illegible]
4	39	6	33	33	37 [illegible]	42 [illegible]
5	38	6	32	32	38	41 [illegible]
6	37	6	31	31	35	40
7	36	6	30	30	34 [illegible]	39 [illegible]
8	35	6	29	29	33 [illegible]	37 [illegible]
9	34	6	28	28	32	35
10	33	6	27	27	30 [illegible]	34
11	32	6	26	26	30 [illegible]	32 [illegible]
12	31	6	25	25	29	31 [illegible]
13	30	6	24	24	27	30
14	29	6	23	23	25 [illegible]	28 [illegible]
15	28	6	22	22	24 [illegible]	[illegible]
16	27	5	21	21	23	23 [illegible]
17	26	5	20	20	21 [illegible]	23
18	25	5	19	19	20	21
19	24	5	18	18	19 [illegible]	20 [illegible]
20	23	5	17	17	18	19
21	22	5	16	16	17	18
22	21	6	15	15	16 [illegible]	17 [illegible]
23	20	6	14	14	15	16
24	19	6	13	13	14 [illegible]	15
25	18	6	12	12	13	14 [illegible]
26	17	6	11	11	12	[illegible]
27	16	6	10	10	11	12 [illegible]
28	15	6	9	9	10	11
29	14	6	8	8	9	10
30	13	6	7	7	8	9
31	12	6	6	6	7	8
32	13	7	[illegible]	5	6	7
33	14	[illegible]	[illegible]	4	[illegible]	6

ILLUSTRATION 4

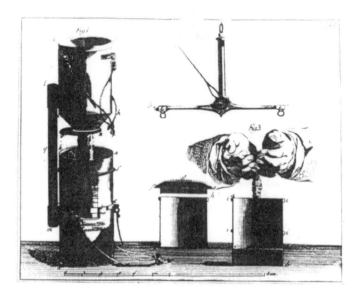

THE SPREAD OF " NEW SCIENCE " TO SOUTHEASTERN EUROPE : DURING OR BEFORE THE GREEK ENLIGHTENMENT ?

Efthymios NICOLAÏDIS

During the 1991 Conference on *New Trends in the Historiography of Science* organized in Corfu, a question frequently aired by the participants was " how did the European phenomenon known as the 'Scientific Revolution' spread ? "

Till recently, historians of science have tried to explain why this phenomenon came about in Europe and not, for example, in the Islamic world during the Damascus or the Tabriz scientific renaissance. Since answers to that question are many and varied, we can safely say that a definitive one has yet to be formulated. This question, along with that of how the phenomenon developed, have been at the fore of historical research into the Scientific Revolution. Very few works, however, have treated the question of how that Revolution spread. A quick perusal of the *Isis* bibliography of the history of science readily verifies this fact : there are extremely few works, and no special category exists on the dissemination of scientific ideas from the centre to the periphery.

In recent years, the European Union has attempted to reinforce its cultural identity by financing scientific projects that promote projects stressing the cultural similarities of the participating countries. One of these, the " Prometheus " project (1994-1996), involved twelve institutes belonging to nine European countries and tried to explain how the ideas of the Scientific Revolution spread from their countries of origin to the European periphery. This question involves European unification through science, through a common European scientific culture. This appears to have been totally accomplished in some countries of the European periphery as late as the second half of the nineteenth century. And here lies the problem which I shall attempt to analyze in this paper. As concerns the countries where the Scientific Revolution originated, we can state that a common scientific culture based on that Revolution had developed by the second half of the seventeenth century.

The activities of Père Marin Mersenne and Isaac Beeckmann sufficiently demonstrate this : the former with his vast correspondence in his capacity of " Secretary " of the *Europe savante*, and the latter with his hospitable role as a *lien vivant* between the European scholars.

When, however, we consider the part of the European periphery constituted by the Greek-speaking world, things appear to be far less clear. At what moment can we say that this world begins to be aware of and participate in the shared culture of European science, formulated by the ideas of the Scientific Revolution ? What were the mechanisms by which these ideas were transmitted to that world ? And, to delineate our main subject, what was the role of the Enlightenment to the dissemination of that science ?

Historical study of the Enlightenment has too often associated the eighteenth century with the spread of the ideas of the new science to the countries which witnessed the elaboration of these ideas (France, Italy, the United Kingdom, Central and part of the Northern Europe). One specific reason explains this tendency : the scientific press of the eighteenth century. From the *Journaux savants* to the *Encyclopédie*, scientific ideas during the eighteenth century circulated widely in printed form. At the same time, these ideas were disseminated to the cultured public in the form of scientific books of a popular nature. This contrasts greatly with the beginning of the seventeenth century, when the priest Jean Tarde had to travel from his native Provence to Italy to obtain a copy of Galileo's book ![1] And what a difference from the researcher of our own day who can study the eighteenth century with a considerable volume of easily consulted printed material, in contrast to the largely archive material of the seventeenth century which for the most part remains unpublished.

In fact, if one confines one's scope to European scholarly society, the ideas of the Scientific Revolution can be said to have spread even before the Enlightenment. The voluminous correspondence of Père Marin Mersenne dating to the 1630s indicates how the new scientific ideas were circulated and discussed[2]. The main difference between the seventeenth and eighteenth century lies in the fact that these ideas were popularized in the latter, and the medium of that popularization was the press.

As concerns the Greek world, a world on the scientific periphery at that period, a glance at the material most readily available (that is books printed in Greek) leads the historian of science to a similar conclusion : that the Greek Enlightenment appears to have provided the venue for the dissemination of the ideas of the Scientific Revolution to the Greek intellectual establishment.

There exist about 140 books containing scientific knowledge that were

1. J. Tarde, *À la rencontre de Galilée : deux voyages en Italie*, Genève, 1984.
2. *Correspondance du P. Marin Mersenne*, Paris, vol. I, 1945 - vol. XVII, 1988.

printed in Greek for the Greek-speaking people of the Ottoman Empire[3]. These date from the invention of the printing press to the time of the Greek Revolution. An analysis made by us of these books indicates that 35 are concerned with knowledge gained during the Scientific Revolution[4]. If we include — as we must — Nikephoros Theotokis in the Greek Enlightenment, only one of these books dates to before this Enlightenment : the book on Spherics and Geography of Chrysanthos Notaras, printed in Paris in 1716[5]. The next book was that on Physics by Theotokis, printed in 1766[6]. The other 33 were printed after 1770.

These statistics are not negated by the publishing boom of the Greek Enlightenment. To mention manuscripts alone, I have found only two manuscript books prior to 1759 that present the knowledge of the Scientific Revolution : the *Physiology* of Vikentios Damodos, of which the oldest manuscript dates to 1738, and the *Epitomé of Astronomy* by Meletios [Michael] Métros, of which the oldest manuscript dates to 1700. On the other hand, since the manuscript tradition lasted throughout all the period of the Greek Enlightenment, we should multiply the 33 printed books mentioned above by a factor of about three if we are to have some idea of the number of the works presenting the knowledge of the Scientific Revolution at this time.

When, therefore, we speak of scientific manuals — printed or in manuscript form — the facts are clear : about 97% of those titles presenting knowledge of the Scientific Revolution to the Greek world up to the time of the Greek revolution date to the period of the Greek Enlightenment.

So much for the statistics and " quantitative data " of the written material. Were one, however, to seek the history of the transmission of these ideas and of the scientific contacts between the Greek world and Western Europe, one would find a significant parameter within which European science was received by Greek scholars during the Greek Enlightenment, from the mid-eighteenth century to the Greek Revolution.

First of all, however, we should try to answer the following question : if one were to exclude the developments of the Scientific Revolution, what, in fact, differentiates the scientific culture of the Greek world in the Ottoman Empire from that of West Europe in the period prior to the Greek Enlightenment ? The dated manuscripts clearly show that it was the legacy of Byzantine science that made that difference. This science was alive and well amongst Greek scholars

3. For the complete catalogue of the Greek printed and manuscript books on science of that period, see Y. Karas, *Οι επιστήμες στην Τουρκοκρατία, χειρόγραφα καί έντυπα,* Athens, vol. I, II, 1993, III, 1994.

4. E. Nicolaïdis, D. Dialetis, E. Athanasiadis, " τυπολογία των βιβλίων των θετικών και φυσικών επιστημών του προεπαναστατικού αιώνα(1700-1821) ", *Τετράδια Εργασίας,* n° 8 (Center of Neaohellenic Research, 1986), 7-38.

5. *Εισαγωγή εις τα γεωγραφικά και σφαιρικά,* Paris, 1716.

6. *Στοιχεία φυσικής...,* Leipzig, vol. I, 1766, vol. II, 1767.

till the mid-eighteenth century. But was this culture so strange to the scholars of Western Europe in the seventeenth century, as is widely believed today ?

Traces of the legacy of Byzantine science in the Greek world prior to the Enlightenment are by no means meager : they consist, on the basis of manuscripts copied during this period, of scientific knowledge formulated during Palaiologean times (1261-1453). This science was far form purely Greek. It represents a mixture of Hellenistic, Persian (the school of Tabriz), Jewish (the Provence Karaits) and Western (e.g. the Alphonsine tables) science. All this knowledge was familiar to European Renaissance scholars, largely from those Byzantine manuscripts which circulated during the fifteenth century, mostly in Italy. This knowledge, however, was not familiar only to Renaissance West-European scholars. As late as 1681, the Astronomer Royal John Flamsteed, an important figure of the Scientific Revolution, presented his students in his famous Gresham College lectures with data drawn from Byzantine astronomers[7]. Furthermore, no one Jesuit astronomer of the second part of the seventeenth and the beginning of the eighteenth century was unacquainted with Riccioli's *Almagestum novum*, where the findings of the Byzantines are presented. Peiresc and Mersenne studied Byzantine music and sought Allatius for information[8]. Western European scholars of the early Enlightenment were, perhaps, much more familiar with Byzantine science than is currently thought. Furthermore, scientific contacts never ceased between the post-Byzantine Greek world and Europe. Even at the end of the sixteenth century, when these contacts were few and far between, Damascene Stoudites adopted the Jewish astronomical tables of the six aisles from a fifteenth-century Byzantine translation, and this work became known in Western Europe[9]. Later on, when the New Science appeared in the West, Greek scholars of the time before the Greek Enlightenment appear to have been much more familiar with the ideas of the Scientific Revolution than our quantitative analysis of the written data would suggest.

One cannot underestimate the role played in pre-Enlightenment Greek science by Chrysanthos Notaras. A nephew of the Patriarch Dositheos of Jerusalem, and later to become Patriarch himself, Notaras dominated Greek scholarly life for more that 30 years. The scientific manuscripts in the Constantinople Annex of the Jerusalem Patriarchate library were mostly collected by him. His book on Spherics and Geography was published in two editions, Paris (1716) and Venice (1718), and was widely quoted in the eighteenth century. Even outside the Greek-speaking world, his influence on the Slave-Greek-Latin Acad-

7. *The Gresham Lectures of John Flamsteed,* ed. by E.G. Forbes, Mansell, 1975, 216-219.

8. Letter of Peiresc to Mersenne, dated 13-15 Oct. 1633, *Correspondance du P. Marin Mersenne,* vol. III, Paris, 1969, 497-505.

9. Παρέκτασις των ιθ' ετηρίδων Μιχαήλ του Χρυσοκόκκου, Library of the Annex in Constantinople of the Patriarchate of Jerusalem, manuscript n° 317.

emy of Moscow was decisive. Later on, as Patriarch of Jerusalem, he had considerable influence over the whole of Greek intellectual and political life.

Let us now consider Chrysanthos' contacts with and his openness to the ideas of the Scientific Revolution. Chrysanthos first studied in Constantinople under Sevastos Kymenites till about 1684. We know that till that time he had not come into any serious contact with the ideas of the new science. Kymenites had himself studied in Italy but we do not have the slightest indication that he was taught anything about this science. Furthermore, the only pre-Enlightenment attempt of great European scholars to present the new science to the Ottoman Empire had failed. This had taken place in 1630, when the new French ambassador to the Sublime Porte planned to have the most famous scholars accompany him to Constan-tinople : Pierre Gassendi and René Descartes were to be in that *compagnie des savants*. Gassendi had even made preparations himself to sail in November of the same year but in the end financial reasons obliged the ambassador to bring only the Hebrew scholar François Galaup de Chasteil d'Aix and Father Theophile Minutti, of the order of the Minimes[10]. After the failure of that attempt, Constantinople had to wait until the Enlightenment before any serious contact could be made with the ideas of the Scientific Revolution, and then only via Greek scholars who had studied in Western Europe. It is interesting to compare this situation with the remote empire of China, where a Jesuit mission of the seventeenth century had included scientists as famous as Adam Shall and Ferdinand Verbiest and had carried a library with the works of Galileo and Kepler, amongst others[11].

To return to Chrysanthos' studies. Without any knowledge of the new science, he was sent on a political and educational mission to Moscow in 1697. Then, for the first time, he came into contact with the new scientific ideas, having obtained from Spathar the Ferdinand Verbiest manuscript which he had brought from his mission to China. In this manuscript, the Jesuit priest and Chief Astronomer of the Emperor Khan-Hi presented the Tsar with the science of the Jesuits. What was Chrysanthos' reaction ? He copied the manuscript at once. Fortunately for us, since the original is now lost[12]. Later on, Chrysanthos studied in Padua where he had few opportunities to learn the new science. Significantly, he furnished his library with a book by Cesare Cremonini, an enemy of Galileo's ideas[13]. Then came his stay in Paris, with its Academy and the

10. See for ex A. Baillet, *La vie de Monsieur Des-Cartes*, Paris, 1691, 228-229.

11. See the more recent bibliography on that subject in N. Golvers, *The Astronomia Europaea of Ferdinand Verbiest, S.J.*, Nettetal, ed. Steyler, 1993.

12. E. Nicolaïdis, " Les Grecs en Russie et les Russes en Chine : le contexte de la copie par Chrysanthos des livres astronomiques perdus de F. Verbiest ", *Archives Internationales d'Histoire des Sciences*, fasc. 133 (1995), 271-308.

13. Ceasaris Cremonini *Explicatio I lib. Meteororum*, Library of the Annex in Constantinople of the Patriarchate of Jerusalem, manuscript n° 210 and *Explicatio II et III lib. Meteororum*, manuscript n° 211. See also the manuscript n° 34, copied by Chrysanthos himself, dated March 27, 1699, when he was in Padua.

Observatory. The fruit of this journey was Chrysanthos' book on Spherics and Geography where he presents the new astronomical systems and — very important — the new mapping methods influenced by Jean Picard. Even though the new astronomical systems are presented with some reservation, this reservation was of a nature typical of his contemporary Jesuits as well. We find here the same reaction as that of many eighteenth-century Greek clerical scholars : in spite of their hostility to the Jesuits, they often adopted Jesuit science and were always interested in Papal reaction to the new scientific ideas. The reason behind this is that these scholars probably felt that the Vatican was secure in its theological Knowledge and that it was prepared to demonstrate or disprove the concordance of any new scientific idea with the Bible. Anyway, Chrysanthos Notaras was, at the turn of the eighteenth century, highly receptive to the new science. The engravings of Descartes' vortexes in the 1716 and 1718 editions were presented to the Greek world for the first time, and this under the aegis of the Patriarch of Jerusalem.

The other great scholar of the pre-Enlightenment period, Methodios Anthrakites, was not so open to the new science. It is perhaps too much to state, as some historians of science have done, that he mastered differential analysis since at that time very few mathematicians in Europe could master this field. Nevertheless, we have convincing proof that, due to his mathematical education in Italy, he was not a stranger to the new methods of analysis[14]. It would appear that, for purely ideological reasons, he chose in his monumental work *Cursus mathematicus*[15] to present Greek students only with a classical education of a nature very close to that of the late Byzantine *quadrivium*. His work recalls that of the Byzantines of the Palaiologan period who remained faithful to the Hellenistic tradition devoid of Persian or other foreign influences. The works of Theodoros Metochites are a good example[16]. Anthrakites' treatise on the astrolabe, included in his mathematical manual, is the last in the history of European science to have drawn from the Greek treatises on the astrolabe based on that by Philopon[17]. Anthrakites, along with many other Greek scholars of the seventeenth and eighteenth century, was a partisan of Renaissance ideas on ancient Greek science. For the Greek scholars, science had been a Greek invention. Greek classical science was the only science that was valid, and the time had arrived to revive this knowledge amongst the descendants of the ancient Greeks[18].

14. M. Lambrou, " Τα μη στοιχειώδη μαθηματικά κατά την εποχή της Τουρκοκρατίας ", *Οι Μαθηματικές επιστήμες στην Τουρκοκρατία*, Athens, 1991, 9-27.

15. *Οδός μαθηματικής...*, vol. I-III, Venice, 1749.

16. For example the monumental work of Metochites, *Στοιχείωσις αστρονομική*, written in 1316 (see A. Tihon, " L'Astronomie byzantine du Vᵉ au XVᵉ siècle ", *Byzantion*, vol. 51 (1981), 603-624.

17. A.Ph. Segonds, *Jean Philopon. Traité de l'astrolabe*, Paris, 1981, 85-86.

18. E. Nicolaïdis, " La tentative de renaissance des mathématiques anciennes dans le monde grec du XVIIIᵉ siècle ", *Études Balkaniques*, n° 2 (1993), 61-72.

In fact, this way of thinking was shared by many Greek scientists of the Enlightenment. In the preface to his book, *Elements of Mathematics*, Nikephoros Theotokis took a similar position — but with a significant difference. According to him, science had indeed been born in Ancient Greece, but it was transmitted to the Europeans, who in turn enriched that knowledge. The modern Greeks need to revive science in their country, this time by taking back the knowledge newly enriched by the Europeans[19]. Here we are confronted with the most significant difference between Anthrakites and Theotokis. In mathematics, the latter attempts a synthesis of ancient geometrical methods and new methods of analysis. His choice to present the mathematics of Gregoire de Saint Vincent is characteristic of his synthetic efforts. Gregoire had attempted to develop a mathematical method to solve problems similar to those solved by Descartes' analysis. This was largely based on classical geometrical methods enriched with the infinites. Theotokis' work *Elements of Mathematics* is a monument to this effort to synthesise ancient Greek mathematics with the infinites, but any such synthesis proved vain in the wake of the development of the tools of differential and integral analysis[20].

Whether partisans or enemies of the new science, the more important Greek scholars wrote and taught science during the pre-Enlightenment or the early Enlightenment period. Notaras, Anthrakites, Theotokis and Voulgaris shared many characteristics. Despite the fact that they were clergymen, three of them of very high rank, they were also men of science. They were university educated in science and mastered the scientific knowledge presented in their writings and teachings.

Some years ago, the present author in collaboration with Demetrius Dialetis attempted an analysis of the main characteristics of Greek scholars who helped to disseminate science in the Greek world between 1700 and the Greek revolution, precisely when the Scientific Revolution was being transmitted there[21]. We attempted to discover information on all these scholars : their education (their studies, the languages they knew, the books they owned), their career, their family, their pupils, their work (scientific or otherwise), their relations with the Church, their engagement in politics, their place of birth, *etc.* This work, published some eight years ago, has now been enriched with the results of the research Projects of the Programme for the History and Philosophy of Science of the National Hellenic Research Foundation[22]. Analysis of the infor-

19. N. Theotokis, " Τοις αναγιγνώσκουσι ", in *Στοιχείων μαθηματικών...*, Moscou, 1798.

20. His method of measure by geometrical methods plus the infinites the volume included betwteen the logarithmic curve and the axes, illustrates perfectly this effort (see *Στοιχείων μαθηματικών...*, 195-198).

21. E. Nicolaïdis, D. Dialetis, " L'influence des Lumières sur la formation scientifique grecque ", *Revue d'Histoire des Sciences*, vol. XLV (1992), 491-499.

22. Website : //www.eie.gr/institutes/kne/ife/

mation about these scholars and their work shows that they fall into three quite clearly differentiated groups categorised by their year of birth.

The first group consists of scholars born before 1750. These people studied at a time when the Enlightenment was not an important element in Greek intellectual life, given that we agree that the Greek Enlightenment begins in the second half of the eighteenth century and becomes a dominant trend after 1774.

A typical representative of this group would be a clergyman who, after having been initially educated in his homeland and eventually at one of the Greek Colleges of the Ottoman Empire, left to study at universities in Italy or perhaps even France. Our student takes courses in Philosophy, Philology, Mathematics and Physics, when the latter are taught separately from Philosophy. Our typical scholar is also highly versed in Theology.

After his studies, our scholar returns, mostly to continental Greece, the Aegean islands or Asia Minor, and takes a position in the clerical establishment of the Orthodox Church, and may teach at a Greek College or privately. Apart from Theology, he teaches Mathematics and Physics and sometimes Natural History. He writes more than three books on Theology, Philology, Philosophy, Mathematics or Physics. His works on science are manuals of a high standard, written in an awkward archaic language. When he has the opportunity to publish a book on science, he sends the manuscript to Italy to be printed.

Both his teaching and his scientific generally concern Mathematics, and Geometry in particular. One need only recall the exalted place of Geometry in the value system of the science of the Ancients (the Byzantine *quadrivium* was headed by Astronomy, but the problem of non-Ptolemaic systems had not really arisen in those times since alternative Arab systems were nothing more than an effort to return to Aristotle's homocenters).

Our paradigm scholar does not aim to present the new science to his pupils. He wants to educate them as fully as possible in classical Greek science, as enriched and formulated by mathematicians such as Tacquet or Saint Vincent. Their teaching is of a high quality ; they do not popularise science and, last but not least, they had been university trained in the science they teach and write about.

The second group of scholars to appear in our analysis was born between 1750 and 1772, at an early stage of Greek Enlightenment. These scholars studied at a time when the Enlightenment was in full swing.

The typical representative of this generation of scholars is no longer a clergyman, but comes from a trading family or is a merchant himself. After a basic education in his homeland, the young student goes to a university in Central Europe (the Austrian empire or the German states), where he studies medicine and/or takes general courses such as geography. He may also learn languages. During his studies, he comes into contact with the ideas of the Scientific Revolution through books or lectures and the discussions and experiments that are very *à la mode* in intellectual circles of the time.

Following this education, our young scholar very often remains and works in Western Europe, mainly in Greek communities, or he may go to the Trans-Danubian Principalities or to Russia, or even to the Ionian islands, which were not controlled by the Ottomans. He is often engaged in revolutionary movements, and will become merchant, doctor or teacher.

Parallel to these activities, our scholar writes some manual on science for Greek students, but also to address this knowledge to a relatively cultivated public. This strategy determines the profile of these manuals : books on Physics or Geography are the most common. They are written in a relatively simple language and their level is relatively low : description prevails over mathematical analysis. The new science is presented frequently here, but almost always in a purely descriptive form. These manuals are in keeping with the popularising scientific books that appear in Europe during the eighteenth century.

It was this generation of " scholars ", obviously, which introduced *en masse* the ideas of the Scientific Revolution to the Greek-speaking world. These ideas were introduced at that time in a descriptive and popularised form : it was the ideas of the Scientific Revolution that were being introduced rather than the actual science of that Revolution.

The next, and last, generation of " scholars " in our analysis was born after 1772, and consisted of men who had grown up at a time when the Greek Enlightenment constituted a dominant intellectual current.

This group is the least homogenous one. Our analysis has included men who wrote books on science, even if at a popular level. Now, however, we find some writers who did not participate in the intellectual life of their time, but only wrote a single popularising book on scientific knowledge. If the scholars of this generation had anything in common, that was their eagerness to communicate some scientific knowledge to the Greek people in order to promote the general level of education. For most of them, promoting this knowledge was synonymous with promoting national independence. It would, therefore, not be an exaggeration to say that many of these scholars wrote books to present the knowledge of the Scientific Revolution to the Greeks for purely political reasons. Let us not forget that the period when these books were written and edited was that of the French and then the Greek Revolution.

An important factor during that generation of " scholars " was represented by the Greek trading communities in West and Central Europe. Just as with the previous generation, merchants played an important role during this period as well.

The fact that this period saw the publication of books which popularised science in general, and the new scientific knowledge in particular, should not mask the fact that the second half of this period saw the appearance of scientific manuals which presented students in Greek colleges with knowledge of the Scientific Revolution in a non-descriptive manner. The *Algebra* by Demetrius Gobdelas[23], the *Philosophie chimique* by Fourcroy[24], and the *Elements of mathematics and physics* by Constantinos Koumas[25] constitute good introduction manuals for those who will continue studies at an university level.

Throughout the transitional period from the classical to the new science, Greek society differed from the countries at the centre of the Scientific Revolution because of the absence of universities. We have catalogued 44 Greek Colleges with an organised educational system that functioned during the seventeenth and eighteenth century, but none of these can compare with the Western European universities of the times. The absence of an organized university system and, more importantly, the absence of Academies during the seventeenth and eighteenth century deprived the Greek world of an organized scientific milieu to act as a recipient of the new scientific ideas. The dissemination of these ideas was, therefore, the result of personal initiatives and any reaction to it could only come from the organized Church. The fact that many of the scholars who transmitted the new science were high-ranking clergymen meant that any reaction from the Church would provoke internal debate, and instances of this kind are known in the eighteenth and early nineteenth centuries. The delayed reaction to the Copernican system made by some clerical circles after the French Revolution constituted an anachronistic attempt to isolate the Greeks from the Scientific Revolution. It included a book by Sergios Makraios in defence of Ptolemy[26] and the publication of the Boulgaris manuscript defending Tycho Brahe[27]. These attempts were hardly echoed — it was too late, in fact, for any reaction of such kind.

Ironically, the absence of universities and Academies led to closer contacts between Greek scholars and their counterparts in Western Europe, since the former were obliged to study at universities at the centre of the Scientific Revolution. While Greeks may not have participated in the formulation of science during the Greek Enlightenment, they were nevertheless in evidence : in Italy, France or Central Europe, they were in contact with the protagonists of that Revolution, they read their books and even translated them. Indeed, the translation of Lalande's *Traité d'astronomie* was made with annotations by the author himself.

Were we directly to answer the question of whether the Greek Enlightenment was the vehicle for the dissemination of the ideas of the Scientific Revo-

23. *Στοιχεία αλγέβρας...*, Halle, 1806.

24. Th. Heliades, *Χημική φιλοσοφία...*, Vienne, 1802.

25. *Σειράς στοιχειώδους των μαθηματικών και φυσικών πραγματειών...*, Vienna, 1807.

26. *Τρόπαιον εκ της Ελλαδικής πανοπλίας κατά των οπαδών του Κοπερνίκου...*, Vienne, 1797.

27. *Περί συστήματος του Παντός...*, Vienna, 1805.

lution, we could answer " yes, but... ". Yes, because the fundamental knowledge of that science spread to the Greek world during the Greek Enlightenment through the teaching of Greek scholars who had studied in West Europe
and through Greek books on science. This dissemination had been well prepared during the pre-Enlightenment period in the Greek-speaking world, from
the end of the seventeenth to the mid-eighteenth century, by scholars of strong
mathematical training, some of whom quite openly presented the main ideas of
the scientific revolution. But this science was also spread during the Greek
Enlightenment for political reasons by scholars who had not studied in the field
of mathematical sciences. During this period, then, the new scientific ideas that
spread to the Greek world were more of a general descriptive nature rather than
involved with the new mathematisation of nature. The inclusion of these new
scientific ideas in the education system was made at the end of the Greek
Enlightenment, just before the Greek revolution. Only after that revolution will
the Greeks begin to participate in the creation of European science thanks to
the foundation of appropriate institutions such as the University of Athens and
the Athens Observatory, where Julius Schmidt formulated his Lunar map. That,
however, is another story, involving different intellectual currents of German
romanticism in science and the French ideal of the engineer-mathematician.

ACKNOWLEDGEMENTS

This paper was given in a first version at a Conference at the " Speros Basil
Vryonis Center for the Study of Hellenism " of the University of Sacramento.
My acknowledgements to Professor Vryonis for his precious remarks.

DISSEMINATION AND DEVELOPMENT OF NON-ARISTOTELIAN PHYSICS IN ARISTOTLE'S LAND

George N. VLAHAKIS

The first question we have to answer is why are we using the term non-aristotelian instead of newtonian to describe physics which was disseminated from Europe to the broader Greek intellectual area (I mean the Balkans, Asia Minor, Propontis, Russia, the Greek communities of Central Europe) during the course of the 18th and early 19th centuries.

Our choice was an outcome of the systematic study of the relevant books, printed or manuscripts, written by the Greek scholars of the era. This study takes place in a long term research project of the National Hellenic Research Foundation, where besides the Greek part, research groups from Bulgaria, Romania, Serbia and Turkey are also participating[1]. The tracing of the new quality and the new identity that the contact with European thought created, and above all, how society as a whole received the new scientific thought, constitute the key reference points of the research. A research which led up to now to a number of general or more specific products (books, articles, the organization of congresses and recently the creation of a valuable database which includes all the available information about the sciences in the Balkans and by no means can be considered just an " electronic attic " of rare and old books).

We have noted that with this effort of the Greek scholars, which is incribed into a wider intellectual movement that of the so-called Neohellenic Enlightenment or better Neohellenic Revival[2], it was not only Newton's system of physics but in parallel the ideas of Kant, Descartes, Christian Wolff and Leibniz which became known to a broader audience, composing a mixture, a com-

---

1. For a brief description of this project see *HSS Newsletter,* vol. 23, 4, Oct. 1994, 6-7.

2. On the general characteristics of this era read : K.Th. Dimaras, *La Grèce au temps des Lumières*, Geneva, 1969 ; G.P. Henderson, *The revival of the Greek thought (1620-1830)*, State University of N. York Press, Albany, 1970 ; P.M. Kitromilides, *The Enlightenment as Social Criticism. Jossipos Moissiodax and the Greek culture in the 18ᵗʰ century,* Princeton, 1992.

mon scientific world, the so-called european scientific thought, which paradoxically enough, seems to be coherent in its basic principles.

The existing differences, between the various systems, substantial for the physicists of Europe, were considered finally as irrelevant for the Greeks who were fascinated by the central idea of the scientific process, the assembling of a scientific method for the determination of the natural laws and the defeat of ignorance and superstition. Ignorance and superstition actually were for the Greek scholars perhaps the greatest obstacle for the popularising of science. An indication for their efforts to defeat these " ugly " characteristics of the society is the translation from the German and the publication in 1810 of the very famous in Europe *Volksnaturlehre zur Dampfung des Aberglaubens*, written by Johahn Heinrich Bertuch.

And what about Aristotle ? Was he expatriated from his own country and got lost in the fog of forgetfulness, simply keeping the renown of an ancient philosopher ? Not at all. Peculiarly, in opposition with what had happened in the West during the scientific revolution, the aristotelian interpretation of the Nature remained an equal paradigm with the european physics. Even in 1816, Sergios Makraios, a learned individual of high esteem and supporter of aristotelianism, published a book called *Epitome of Physics* where he tried to compromise Aristotelian physical philosophy with contemporary physics[3].

The inconsistencies and the mistakes were yielded to the medieval commentators of the aristotelian work.

In the Balkans Aristotle's thought was refined, for the problematic parts of his texts, through the very significant for the period work of the Greek neoaristotelian philosopher Theophilus Korydaleus, who had excellent studies under the wellknown Italian professor Cesare Cremonini.

Aristotelism remained present at least until the establishment of the independent Greek state, when through other procedures contemporary physics prevailed completely, but this is not our subject here. We shall speak about the non-aristotelian physics. For the convenience of our European colleagues we present a short description of the books circulated during the period under examination.

At the beginning of the 18th century, in 1738, Vikentios Damodos, a wellknown scholar among the Greek intellectuals, wrote *Physiologia* or General Physiology, a work which remained unprinted, but spreaded widely and had significant influence during the first half of the century. Damodos seems to support Descartes opinions, but he refers to Newton's theories first of everybody else in the Greek area.

---

3. On a reappraisal of the role of S. Makraios see : G.N. Vlahakis, The " other view " : Sergios Makraios's *The Epitome of Physics,* in : *The Sciences in the Greek area,* Athens, Centre of Neohellenic Research, NHRF, 1997, 249-260 (in Greek).

Some very foundamental ideas of the newtonian physics would be found also later in the Greek edition of Rollin's *Ancient History*, translated by Alexandros Kangellarios and published during 1750.

But actually the turning point was the publication of a book written by Nikephoros Theotokis, perhaps one of the very few persons who one could characterize not only a scholar but a scientist too. The book entitled *Stichia Physikis* (Elements of Physics) was published in Leipzig in the years 1766-1767[4]. We consider it as the most important book of Physics published during the 18th century because for the first time newtonian thought becomes acceptable in the Balkans.

As we are going towards the end of the century, subjects of physics are treated in the popularising books *Florilegium of Physics* written by Rigas Velestinlis and *Grammar of the philosophical sciences* written by Anthimos Gazis.

Gazis's book was based on the Italian edition of the *Grammar of philosophical sciences*, an introductory book to newtonian physics written by Benjamin Martin, an English instrument-maker and peripatetic lecturer. Two other books were published during the first years of the nineteenth century though written by Eugenios Vulgaris about two or three decades ago. These books were *Ta areskonta tois philosophois* (*Principles of the philosophers*) and *Peri systimatos tou pantos* (*About the system of the universe*). In 1812 three books of Physics are published, D.N. Darbaris's *Epitome of Physics*, K.M. Koumas's *Synopsis Physikis* (*Concise Physics*) and Konstantinos Vardalahos's *Experimental Physics*.

Late works of the period are the *Physics* of Benjamin Lesvios and the *Physics* of Theophilos Kairis, which finally remained in a manuscript form, but had as the first work we have mentioned, that one of Damodos, a significant impulse on the science of the era. Both Lesvios and Kairis supported the existence of a matter of aethcrial character, which the first one called *Pantahikiniton* (a substance which is moving everywhere) and the second called *Enylon*.

This production small in absolute numbers, small compared with the relevant numbers in Europe, shall not be considered negligible according to the social, financial, political and cultural status of the society in the region.

Two words concerning the way Greek scholars became acquainted with Physical sciences. The junction point was Italy. There, wealthy Greek communities of merchants had been established in cities like Trieste, Venice and Livorno. Because of this situation young Greeks, especially from Crete and the Ionian islands, had the privilege to study in high level schools like the Cottunian and the Flaginian colleges and afterwards to attend lessons in the Univer-

4. G.N. Vlahakis, *The Physics of Nikephoros Theotokis : a turning point in the scientific thought of the 18th century,* Ph.D. Thesis, Athens, National Technical University, 1991 (in Greek).

sities of Padua and Bologna at the beginning and in the Universities of Pavia and Pisa later[5].

It is the University of Padua where, by good chance, four of the introducers of the contemporary scientific world, Nikephoros Theotokis, Eugene Vulgaris, Nikolaos Zerzoulis and Iossipos Moissiodax studied. They studied exactly at the period when the great reformation at this University took place under the guidance of the famous Italian professor Giovanni Poleni who degraded the aristotelian philosophy and established the teaching of the experimental Physics in the wellknown *Teatro sperimentale*[6].

The influence of Poleni had acted as a kind of catalyst to the young Greek students who were refering to him in their writings with respect and admiration.

But there was also a need for a continuous written reference to whatever they had taught relative to Physics and their multiple activities after their return to Greece should blunt. Peter van Musschenbroek's *Elementa Physicae* in principal and secondary Abbe Nollet's *Lezioni di Fisica sperimentale* in their Italian editions served as this kind of references[7]. The use at the same time of the opinions of the popularizer of the newtonian work in Europe, Peter van Musschenbroek, with those of the cartesian founder of a special theory of electricity Abbe Nollet, are compiled in a way which could be characterized as the " Hellenic synthesis ". One could for a moment consider as exaggerated the expression Hellenic synthesis. But the detailed study and analysis of the texts written during this period could persuade even the most distrustful that this is not far from the truth. Undoubtly, one who wants to deal with History of Science and not with Mythology of Ideas, owes to check thoroughly the sources and the evidence which are available, avoiding to express speculations in an effort to transfering the schemes used for the determination of the science's conditions in Europe. That is because we deal with what happened not in the centre, where science is produced, but in a periphery where science is disseminated. In a periphery where contemporary physics is connected directly with two not scientific visions. The national revival and the freedom from the Ottoman occupation.

It is not accidental that there is not even one book of that period where the achievements of the Western thought since the scientific revolution were considered as an indication of overthrow of the ancient Greek thought. Instead, the new physical theories are taken as expressions through a contemporary formal-

5. G.N. Vlahakis, " An outline of the introduction of classical physics in Greece. The role of the Italian Universities and publications ", *History of Universities* (1998), 157-180.

6. G.A. Salandin and M. Pancino, *Il Teatro di filosofia sperimentale di Giovanni Poleni,* Lint, Trieste, 1987.

7. G.N. Vlachakis, " L'oeuvre scientifique de Niképhoros Théotokis : tentative d'approche fondée plus particulièrement sur les *Stichia Physikis* (*Eléments de Physique*), *RESEE,* XXV, 3 (1987), 251-261.

ism of alternative philosophical theories of the ancient world, where Aristotelism keeps its validity at least in the part relative to methodology.

Lets cite just one example from Pamblekis : " It seems to me that the germs of the teaching of the famous Leibniz and the students of Wolff, could very easily be traced in the theories of the ancients. Because Leibniz's monads and Wolff's elements what difference have with the ideas of Plato and Pythagora's immaterial principles. Leibniz himself admits, that he borrowed from Plato, because he found in his opinions truths not existing elsewhere ".

Before a while we talked about hellenic synthesis and we supported that it actually realised. Indeed, Daniel Philippides, in early 19th century, wrote an *Elementary Treatise on Physics* which remained unpublished. We will not proceed to an extensive analysis of its content. We shall refer only to the preamble which led the author to undertake the writing of such a book : " But after that period, which one may consider as a true renewal in the observational sciences, the findings published each one separatedly. They are found in the proceedings of the various learned societies and in some special treatises and noone has tried yet to join them in a body of learning. We missed therefore a treatise of Physics, where things shall be arranged according to their mutual dependence, to a small number of general phenomena, which one may consider as principles, and there these principles to be arranged systematically ".

This foundamental methodological approach for writing a textbook of Physics I think that it is even today in fashion. But even more radical is another idea read in the same text : " I have not accepted any system. I think that they are often responsible for the interruption of the progress of physics and they are quite harmful for the sciences in general. To create the systems, people do hypotheses without any background and usually incomprehensible, and after repeating these hypotheses for 10 or 12 times they think that they are proved adequately ".

This antidogmatic view which free the spirit from the barren connection to ideological schemes which attempt to achieve a science-like appearance is compiled completely with the only approach of the nature approved by the " enlightened " Greek scholars of the period, the experiment.

It is not a fact of mere accident that the experiments, either when just described or when performed, according to the sources we have, are treated as a practical confirmation of the right thing and not as a contradiction of the mistaken having also always a beneficial-educational component[8].

On the other hand experiment served also as a procedure for the development of the self-control and the selfdiscipline necessary to exist in a nation under formation, a nation which for centuries was used to act under a kind of

---

8. About the role of the experiments see : C. Xenakis, *The experiment as method of research and knowledge in the works of the scholars during the prerevolutionary period*, Ph.D. Thesis, Ioannina, 1994.

anarchic nationalism, though there were some structures of authority both by the Turks and the circles around the Patriarchate.

Let us return to the examination of the ways used by the intellectual circles of the 18th century to learn physics. Besides the very important role of the Italian Universities and publications, at the turning of the century two other centres of great influence were arised. France and the German speaking countries. The French Enlightenment and the German *Aufklärung* played a role for this fact. Not only philosophy in general but sciences as well and physics in particular have been disseminated in Balkans from these countries[9].

To restrict ourselves in the case of Physics we shall mention that from the books we have noted previously, two have as basic source the French *Encyclopédie* of d'Alembert and Diderot. But the Greek authors used it with entirely different way.

Christodoulos Pamplekis in his book *On Philosopher, Philosophy etc.* aimed to construct a text discussing the materialistic principles of contemporary mechanistic physical philosophy, while Rigas Velestinlis tried to write simply an educational elementary text for students, who for the first time learn what is Physics and which is its subject.

Both, Pamplekis and Rigas succeeded proving this way how important was the influence of the *Encyclopédie* but also how capable were the Greek scholars of the era to exceed the phase of the simple translation and conquer higher stages. The Greeks usually remoulded, reformed and enriched their material and therefore they had the right to be considered not mere translators but producers of scientific work.

Not only *Encyclopédie* but other French books of Physics were also used as sources. For example Pierre Adet's *Leçons élémentaires de Physique* served as the original of K.M. Koumas's *Concise Physics.*

Koumas is in parallel the individual who accepted the most intense influence from the German area. His basic book of philosophy was based on Krug's work while in his Physics he used also Gren's textbook of Physics. Konstantinos Vardalahos and D.N. Darbaris in their books of Physics accepted also the influences of the German area.

All these books have undoubtly a newtonian character. But there are subjects where we find also non newtonian theses, mainly these which give a possibility for philosophical or even metaphysical discussion. For example the nature of the physical body, what are the fundamental particles of the matter, the nature of the forces, the existence and the properties of the vacuum and aether[10], what we mean by space and time *etc.*

9. See : G.N. Vlahakis, " A note for the penetration of Newtonian scientific thought in Greece ", *Nuncius,* 2 (1993), 645-656.

10. See : G.N. Vlahakis, " Philosophical and Scientific arguments on the existence of vacuum. The case of the Greek scholars during the XVIIIth century ", *Phlogiston,* vol. 3, 6 (1997), 73-95.

Of particular interest was also the discussion of the nature of light and electricity, which constituted a field of scientific debate also in Europe[11].

In that case Greeks followed usually the middle way. They presented the proposed theories avoiding to support either of them. But not always. Sometimes as for example in a very interesting article published in *Hermes the Scholar*, the only journal with scientific contents in the Balkans during the early 19th century, we read that the author prefers the materialistic theory of Newton than Euler's wave theory.

We promised to present also some examples of the way Greeks incorporated the non-aristotelian physics by the Greek scholars. Let's start from Vikentios Damodos comments on the subject of the nature of the physical body : " About the physical body very famous is the opinion expressed by Descartes and his disciples. On the other hand the Newtonians reject that theory as false unanimously, using theological arguments ".

This short description of some characteristics of the scientific practice in the Greek intellectual area permit us to express a scheme for the dissemination of the scientific thought from the centre to a periphery. According to our opinion this dissemination, as for the quantity but also as for the quality, is not depending solely on the produced science in the centre, but on factors, usually non scientific prevailing in the periphery. For example in a country of colonial type, it is obvious that the metropolitan oriented ideas will be accepted rather automatically. That happened in Greece after the establishment of the independent state, when actually developed a climate of subjection in the developed countries like France and Germany.

On the contrary during the period of the Ottoman occupation scientific knowledge used also as a mean for the formation of the national consciousness resulting in the existence of reflections and refractions owed to ideological criteria, refractions which we outlined here. Furthermore we must not forget that in our case the periphery is characterized by a transient nature. Though it was eastern in many dimensions on the other hand the intellectual vanguard wanted desperatedly to incorporate it in the West as the prevailing opinion was that Western culture was more or less the result of the ancient and byzantine influences. Besides the political and social factors the financial status imposed rectrictions to the dissemination of science.

The poor financial means did not leave any margin for the development of the structures which would permit the basic or applied research.

Consequently we should not criticise negatively the choices of the Greek scholars without previously studying thoroughly their work in relation to the

11. J.L. Heilbron, *Electricity in 17th and 18th centuries. A study of early modern Physics,* University of Callifornia Press, 18, and for the case of the Balkans : G.N. Vlahakis, " The Reception of the Electric Theories during the Dawn of the Neohellenic Scientific Thought ", *Nuncius,* 2 (1998), 87-106.

particular conditions of the Balkans during the 18th century and that is what we try to do in our project.

LA MÉTHODE EXPÉRIMENTALE EN GRÈCE
À LA FIN DU XVIIIᵉ SIÈCLE : INFLUENCES ET RÉACTIONS

Christos T. XÉNAKIS

INTRODUCTION

Pour bien pouvoir comprendre l'expansion et l'assimilation de la nouvelle pensée physique et sa méthodologie dans l'espace intellectuel grec, à la fin du XVIIIᵉ siècle, il faut d'abord voir brièvement la situation déjà existante dans le domaine des sciences physiques, son évolution et ses liens avec le vaste patrimoine culturel hellénique.

Les textes de physique, imprimés ou manuscrits, concernant la période qui nous intéresse, expriment l'idée que la philosophie est étroitement liée aux sciences physiques, si bien que le développement de la philosophie suggère celui, parallèle des sciences physiques.

Une étude des sources et de la bibliographie nous permet de constater que, jusqu'en 1750 et même un peu plus tard, jusqu'à la première apparition de textes imprimés portant sur des notions et des théories physiques formées en Europe, cette science n'est qu'une considération philosophique de la nature c'est-à-dire une philosophie physique et elle s'identifie à la physiologie d'Aristote[1].

Le cadre dans lequel s'est développée la physique grecque dans la première moitié du XVIIIᵉ siècle se définit par différents textes, simples de conception, qui essayent d'interpréter les phénomènes physiques à l'aide des textes ecclésiastiques et des superstitions. La peur, ainsi cultivée, certifiait la dépendance directe de l'homme vis-à-vis de son environnement naturel[2].

Ce cadre peut également être défini, pour la période historique qui nous intéresse, par la tradition aristotélicienne et l'aristotélisme qui a constitué,

---

1. V. Damodos, *Φυσιολογία αιτιολογική εις την κοινήν διάλεκτον, αριστοτελική, σχολαστική και νεωτερική*, code n° 67, Samos, Grèce.

2. *Ερμής ο Λόγιος*, 1818, 108.

comme ensemble de connaissances et de méthodologie, une base pour l'évolution de la pensée néohellénique.

L'étude de la bibliographie imprimée ou manuscrite antérieure à 1750, permet de constater que l'étude des sciences physiques, est alors surtout fondée
sur le livre d'Aristote intitulé *Physsikis Acroassis* commenté principalement
par Théophilos Korydaléas. Celui-ci nous délivre la pensée d'Aristote pure et
dégagée de toutes les interprétations arbitraires issues des milieux religieux et
des falsifications du Moyen Age[3]. L'observation et l'étude des éléments qui
constituent le monde naturel, la classification des phénomènes naturels dans le
but de démontrer l'harmonie du Cosmos, l'utilisation de la méthode inductive-
déductive dans la recherche, sont quelques-unes des connaissances dont s'enri-
chit le savant grec par son contact avec l'aristotélisme.

Mais surtout l'étude systématique de la pensée aristotélicienne a suscité,
chez le savant grec, un grand intérêt pour les sciences physiques et les problè-
mes du monde naturel. Elle lui a offert la satisfaction de pouvoir suivre, dès
ses prémices, la théorie de l'évolution de la nature, en lui permettant de saisir
l'importance du mouvement physique, responsable de tout phénomène naturel.

L'ACCUEIL DE LA NOUVELLE PENSÉE PHYSIQUE
ET DE LA NOUVELLE MÉTHODOLOGIE

En Italie, pendant la même période, des étudiants grecs se trouvaient à
Padoue. Pour la plupart, ce sont des disciples d'Aristote, et parmi eux, certains
adoptent la nouvelle physique et son orientation vers l'expérience et les mathématiques. Quelques décennies plus tard Nicolaos Zerzoulis, en traduisant une
partie de l'oeuvre de Newton[4] déclare adhérer à la nouvelle physique et à la
nouvelle cosmologie.

Pendant le dernier quart du XVIIIe siècle, les principaux représentants grecs
des " Lumières " acceptent le système héliocentrique. Leur choix est la consé-
quence directe de l'adoption générale du contenu de la nouvelle physique et de
la méthode expérimentale.

Lossypos Moesiodax constitue un exemple typiquement avant-gardiste de
ce mouvement : l'étude de ses oeuvres montre qu'il a bien conscience que la
nouvelle physique ne se contente pas d'améliorer ou de réfuter certaines idées
de la philosophie aristotélicienne, mais qu'elle va beaucoup plus loin. Elle
remplace le modèle du monde traditionnel, l'univers restreint, fermé et hiérarchisé par un autre, nouveau, qui nous oblige à reconsidérer radicalement la
place de l'homme dans le monde.

3. N. Psimmenos, *Η Ελληνική Φιλοσοφία*, éd Gnosi, 1988, t. 1, 183.
4. I. Newton, *Philosophiae Naturalis Principia Mathematica*, Londini, Anno NMCLXXXVII.

C'est dans ce climat de contestation que l'esprit européen et les nouvelles idées s'inscrivent dans la pensée scientifique grecque, jusqu'alors encore très influencée par l'aristotélisme.

Parallèlement, l'ancienne pensée grecque (antique et byzantine) survit et alimente la pensée néohellénique. A la fin du XVIIIᵉ siècle la pensée physique grecque évolue sur un axe bipolaire : " tradition-renouvellement ". Surtout en physique, le passage est très clair d'une pensée " empirique-descriptive " à la pensée " analytique-expérimentale ", c'est-à-dire de la simple observation et description des phénomènes à l'aide des mathématiques et des expériences[5].

Les savants grecs qui étudient en Europe au XVIIIᵉ siècle, connaissent la méthodologie expérimentale et, de retour en Grèce, ils l'enseignent aux autres. Dans leurs textes sur les sciences physiques, ils se réfèrent à la méthode expérimentale comme à une méthode de recherche et de connaissance du monde naturel. Pourtant, ils ne se réfèrent à l'expérience que pour citer celles des grands chercheurs des XVIIᵉ et XVIIIᵉ siècles comme Galilée, Torricelli, Boyle, Pascal, Newton, Laplace, Coulomb, *etc*. Leurs expériences sont cependant décrites sans détails et bien sûr sans aucune possibilité de reproduction, les seuls " instruments " dont ils disposent se réduisent à quelques schémas hors texte[6].

LA RECHERCHE

L'étude que nous avons entreprise est réalisée dans le cadre du programme de l'Institut de Recherches Néohelléniques de la Fondation Nationale des Recherches Scientifiques : " Influences de la pensée scientifique européenne dans l'espace grec et balkanique aux XVIIIᵉ et XIXᵉ siècles ". Son but est de rechercher, recueillir et exploiter les influences et les réactions provoquées par l'introduction de la méthode expérimentale comme méthode de recherche et de connaissance dans le monde intellectuel grec.

Dans ce but, ont été consultés tous les livres imprimés ou manuscrits qui se réfèrent aux sciences physiques de l'époque, la correspondance entre savants, du XVIIIᵉ et du début du XIXᵉ siècle ainsi que toute autre source, grecque ou étrangère qui pourrait fournir des informations sur l'introduction de la méthode expérimentale dans l'espace grec (journaux, revues, annales, *etc*.). Ainsi, sommes-nous parvenus à créer une base de données qui nous a amenés à de premières conclusions sur les influences de la nouvelle méthodologie dans la recherche et la connaissance du monde naturel, le changement des mentalités

5. J. Karras, *Οι Θετικές επιστήμες στον ελληνικό χώρο, 15ᵒˢ-19ᵒˢ αιώνας*, Athènes, éd. Zacharopoulos, 1991, 130.
6. C. Xénakis, *Το πείραμα ως μέθοδος έρευνας και γνώσης στα έργα των λογίων της προεπαναστατικής περιόδου*, Ph. D. Thesis, Jannina, 1994, 147.

et en général sur les idées que les Grecs adoptaient pour interpréter les phénomènes naturels et combattre les superstitions.

INFLUENCES ET RÉACTIONS : DISCUSSION ET CONCLUSION

Les nouvelles idées sur les lois et le fonctionnement du monde ainsi que les
méthodes utilisées (méthode expérimentale, méthode déductive-inductive à
l'aide des mathématiques) ont été fortement contestées par les cercles conservateurs.

Cette contestation porte surtout sur l'expérience, parce qu'elle assure une
connaissance objective et par conséquent qu'elle ébranle l'édifice religieux et
social très hiérarchisé tout en renforçant l'esprit critique et les arguments des
rationalistes et des empiriques.

D'après l'étude faite, l'accueil de l'expérience, dans le cadre d'une nouvelle
méthodologie et philosophie physique, fut, du moins au début, négatif. On con-
testa fort que l'expérience puisse aider à comprendre les phénomènes naturels
de façon aussi convaincante que l'aristotélisme.

D'ailleurs, il ne faut pas oublier qu'à cette époque en Grèce une réaction
générale s'était développée contre l'esprit occidental qui diffusait les idées de
la Révolution Française et le matérialisme européen. Le siège patriarcal et la
société bourgeoise conservatrice ont fortement réagi contre une introduction
possible de la nouvelle méthodologie dans les programmes scolaires.

A notre avis cette réaction résulte aussi de la peur de ces couches sociales
devant l'influence, de plus en plus visible, du catholicisme et du protestan-
tisme. Les missions de ces églises dans les territoires de l'empire ottoman ont
effarouché les cercles conservateurs de l'église orthodoxe qui ont eu le réflexe
de se retrancher derrière des thèses aristotéliciennes. A l'opposé de cette réaction, une nouvelle méthode d'acquisition de la connaissance se développa, surtout d'après la théorie de Locke[7], qui utilisait l'expérience pour réévaluer la
nature et qui s'opposa à la théologie grecque. L'utilisation de l'expérience pour
la vérification de la relation " cause-effet " dépouilla la nature de tous ses mys-
tères et forces extraordinaires, tout en la rendant responsable de tous les phénomènes naturels.

Plus tard, on constate un changement favorable d'attitude envers la nouvelle
méthodologie. Le revirement est dû à l'amélioration du niveau de vie de la
bourgeoisie moyenne, à sa prospérité économique obtenue grâce au commerce,
à son ascension sociale et surtout à ses contacts financiers et culturels noués
avec l'Europe. Ceux-ci ont nécessité une meilleure éducation afin de permettre
la compréhension des phénomènes naturels et de la technologie et non plus la
spéculation métaphysique. Le besoin de comprendre le " pourquoi " et le

---

7. J. Locke, *An Essay concerning Human Understanding*, (1689) voir réimpr. Oxford, 1924,
repr. 1969.

" comment " du monde qui nous entoure favorisa l'adoption de la nouvelle méthodologie comme méthode de recherche et de connaissance[8]. Aussi, ce changement est-il favorisé par l'objectivité de la connaissance fournie par l'expérience, la commodité de la nouvelle méthode et l'approche critique des théories physiques déjà existantes par les savants de cette société.

D'après l'étude des sources, il apparaît que l'expérience a été utilisée par les savants grecs comme moyen d'observation, de reproduction et de compréhension des phénomènes physiques, sans pourtant disposer de mesures exactes qui permettraient de vérifier certaines lois. Elle a été utilisée pour soutenir plusieurs théories physiques sur la nature de l'électricité, de la chaleur, de la lumière et du son.

De même, l'expérience a fonctionné, directement ou non, comme une méthodologie de recherche des causes des phénomènes naturels mais aussi comme une méthodologie révélatrice de la profonde relation de la cause et de l'effet. L'étude des oeuvres des savants grecs de la fin du XVIIIᵉ siècle nous montre que l'expérience a aidé la connaissance scientifique à devenir une application pratique et à s'affirmer au moyen d'observations, de répétitions et de procédures incontestables.

Enfin, l'expérience a joué un rôle important dans la réévaluation de l'enseignement des sciences physiques : dès le début du XIXᵉ siècle, les écoles grecques de l'empire ottoman se sont équipées de laboratoires (cabinet de physique) où se pratiquent des expériences de mécanique, de changement de la température, d'acoustique, de magnétisme, d'électricité et quelques expériences de chimie, mais toujours au niveau de simple démonstration.

En conclusion, il est évident que l'expérience a été, pour les savants grecs un " outil " méthodologique pour étudier et comprendre le monde naturel. Il ne fut plus nécessaire de recourir à des hypothèses exagérées ou à des forces secrètes pour expliquer les phénomènes naturels. Il suffit désormais de faire confiance à la raison, au pouvoir de la mécanique de Newton et à l'exploitation de la technologie à l'aide d'expériences de démonstration bien organisées.

---

8. *The Uses of experiment*, ed. D. Gooding, T. Pinch, S. Schaffer, Cambridge, Cambridge University Press, 1989.

The Spread of the Scientific Revolution into Russia
The Rise of Physics in Russia

A.A. Pechenkin, D.L. Saprykin

Three attempts to establish physics in Russia

The response of the Russian culture in the 17th and 18th centuries to the challenge of the Scientific Revolution (more precisely the lack of such a response) is a kind of enigma.

First, there seemed to be a favorable ground for the expansion of Modern European science into Russia. Indeed, as early as the 17th century there was a rather high level of elementary and secondary education in Russia. According to a widespread opinion, it was the reforms of Peter the Great that were a stimulant to the growth in education in Russia. In contrast to this, A.I. Sobolevsky, the outstanding Russian historian and philologist, considered that elementary education tended to decline after those reforms since the tradition of home and church education was destroyed in this process[1].

Second, there were advanced engineering and skills in Russia at that time. In the 17th century, the flowering of the Russian architecture, the rise of the so-called Moscow style was in progress. Besides architecture, the very complex and complicated principles of town-planning and fortification had developed. For instance the fortification and life-support system of the Solovetskii Monastery provides evidence of the high level of engineering and applied mechanics.

Third, there is evidence that representatives of high society were well-educated. For example, St. Metropolitan Gennadiy (Gennadiy Novgorodsky), who is not usually regarded as the top Russian intellectual of that time, initiated and ran the quest for the first complete Slavic codex of the Bible which was issued in 1499. He was also trained enough in mathematics to perform the rather

---

1. A.I. Sobolevsky, *Obrazovanie v Moskovskoi Rusi* (Education in the Moscow Russia in the 15th-17th Centuries), Saint-Petersburg, 1892.

complicated calculations of the " Great Indiction " (*Circulus Magnus*) in connection with the problem when Easter should be celebrated over the " eighth millennium since the Creation " (after 1491)[2].

The above fortification and life-support system of the Solovetzkii Monastery was erected under the guidance of Abott St. Philip (afterwards Moscow Metropolitan), who was a friend of Tsar Ivan IV but subsequently became his spiritual enemy and was killed by order of the latter. In his memoirs, written at the end of his life, St. Philipp presented arguments against Ivan's rule and a verbal portrait of the bad tsar.

In the second half of the 17th century, among Russian clergymen who taught at ecclesiastical schools, especially at the Kiev-Mogila Academy and the Slavonic-Greek-Latin Academy (Moscow) there were people who had studied in the best universities of Europe : St. Maxim Greek, the Leichoudi brothers, Adam Zoernicav, Palladiy Rogovskii, *et al.*[3]. Epiphanii Slavinetskii, Head of the school at the monastery of Chudov and the rigorous orthodox theologian, leader of the anti-Latin party in Moscow, for example, was well-educated in mathematics and astronomy and his erudition in the ancient languages and philology was eminent. At the end of the 17th century, students of the Slavonic-Greek-Latin Academy studied the full course of physics, which had mainly been borrowed from Italian universities[4].

The level of education of statesmen should also be considered. The outstanding statesman of the period of tsar Alexsey, father of Peter the Great, A.L. Ordin-Nashekin, who could speak German, Polish, Latin, was educated in mathematics. Prince V.V. Golitsin, the Noble in attendance on Tsarina Sofia, Peter's sister, of whom he bereaved power, read German and Polish books, was acquainted with elements of geometry, astronomy, geography[5].

After all, the attitude of the Russian state and Church to modern science was very favorable (especially under Peter the Great and Catherine the Second). The Saint Petersburg Academy of Sciences founded by Peter the Great was very attractive and prestigious, and many outstanding European scholars enrolled at this Academy.

Nevertheless, modern science (first of all physics), generated by the Scientific Revolution, had not set on Russian grounds for a long time. If a " scientist " means a modern European scholar, one can agree with the Russian physicist P.P. Lazarev, who stated that " before Peter the Great, Russia had no

2. When the Great Indiction (532 years) elapses the Easter full moon falls back to the same day and date.

3. A.V. Kartashev, *Ocherki po istorii russkoi tserkvi* (Essays on the History of the Russian Church), Moscow, Terra, 1992.

4. G. Mircovich, " O shkolah i prosveshenii v patriarshii period " (About Schools and Education in the Patriarch's Period), *Jurnal Ministerstva Narodnogo Prosviashenia* (Journal of the Ministry of Public Education), 7 (1978).

5. V.O. Kluchevsky, *Istoricheskie portrety* (Historical Portraits), Moscow, izd.Pravda, 1990.

scientists whatsoever". Science imported into Russia by Peter the Great, " science by decree ", as Vucinich calls it[6], failed to stand on its own Russian feet. No Russian schools followed the great Saint Petersburg physicists of the 18th century : Leonard Euler, Daniel Bernoulli, Franz Aepinus, and Michail Vasilievich Lomonosov.

The Russian historians of science experienced great difficulties as they aimed to trace a line between the Great Saint Petersburg academicians Euler and Daniel Bernoulli and the Russian physics of the 19th century. They have to adjust their observations, and speak too generally and vaguely.

No school followed the Great Russian academician M.V. Lomonosov, who had passed through Slavonic-Greek-Latin Academy and had been educated as a physicist in Germany. It should be noted that Lomonosov was most interested in electricity and heat, the fields that had been theoretically developed over the 19th century.

The history culminates in the impression that Russians were not simply interested in general mechanical physics which followed Galileo, Descartes, Huygens, and Newton.

There were three attempts to develop physics in Russia, the third being only actually successful. The first took place under Peter the Great and the following Empresses.

Physics were imported again into Russia in the third decade of the 19th century from German-language countries. Those were the formative years of the Saint-Petersburg academicians E.Ch. Lentz, B.S. Yakobi, A.T. Kupffer, et al., who worked at the department of physics of the Academy of Sciences. Emil Christaforovich Lentz (1804-1865) graduated from the University of Dorpat and began to work at the Saint-Peterburg Academy of Sciences in 1828. Boris Semenovich (Moritz Herman) Jakobi (1801-1874), who graduated from the University of Goettingen and worked at the University of Koenigsberg, came to Dorpat University as professor of physics in 1835 and was soon invited to Saint-Peterburg where he became an Academician in 1839. Adolf T. Kupfer (1799-1865) graduated from the University of Berlin, became professor at the University of Kazan in 1823 and a member of the Saint-Petersburg Academy in 1828.

Since Lentz, the Russian educational tradition in physics had been notified. Lentz had an impact on higher education in physics as he accepted the chair of physics at the University of Saint Petersburg (the mid-1830s), became Dean of the Physico-Mathematical Faculty in 1841 and Rector of the University in 1862. Lentz's disciples, F.F. Petrushevsky, M.P. Avenarius, F.N. Shvedov played their part in the development of the education in natural science in Russia.

6. A. Vucinich, *Science in Russian Culture. A History to 1860*, Stanford Univ. Press, 1963, 38.

Although the famous Russian physicists of the end of the 19th century D.I. Mendeleev (1834-1907), A.G. Stoletov (1839-1896) and N.A. Umov (1846-1916) spent some time in German Universities, they graduated from Russian Higher Schools and developed physics in a Russian way, that is within the framework of applied problems and with a special emphasis on experiment.

A.G. Stoletov and N.A. Umov together with P.N. Lebedev (1866-1912), who were educated as physicists in Germany, were representatives of the third wave of Russian physics. Here, the steady tradition, not only in education, but also in research had been notified. Although Lentz, Jacoby, Kupffer greatly contributed to Russian physics, they basically achieved progressive shifts in education. However, advanced education depends on advanced research. As a result in the 1860s Russian university, physics was well apart from the European level. As N.A. Gezehus recalls, physics in Russian universities was " in childhood and complete stagnation. There were no spots of scientific schools, students had no practical exercises, without which it is difficult to educate students as good physicists. There were no magazines and societies that helped in discussing new ideas and stimulated scientific activity "[7].

In Russian universities, special emphasis was placed on the humanities and mathematics, but not on physics.It is perhaps for this reason that, by the second half of the previous century, Russia had had great literature and there had been steady traditions in mathematics in Russian Universities. Russian physics could not yet keep up with the time.

In Russia, the steady growth of physics started because of endeavours of Stoletov, Umov and Lebedev at the University of Moscow and Mendeleev and Petrushevsky at the University of Petersburg. At the beginning of the 1870s, Stoletov set up the physical laboratory and practical work for students, basing himself on his experience in research at the University of Heidelberg. Lebedev who started as a physicist at the University of Strabourg (1887-1888) and worked at the University of Berlin and after that at Strasbourg again (1890-1891), played a decisive part in organizing the Institute of Physics at the University of Moscow. In his Moscow period, Lebedev was very influenced by Stoletov and cooperated with him. By the end of the 1860s, a physical laboratory had been established at the University of Petersburg but in Moscow, physics was more advanced.

It should be emphasized that European science had strongly changed by the beginning of the 19th century. Roughly speaking, it became rather positivist and distanced itself from or repudiated metaphysical problems. It was influenced by a pragmatic orientation of the French scientists and descriptive tendencies

7. N.A. Gezehus, " Historical Outline of the Physical Society at Imperial Saint Petersburg University ", *Journal russkogo fisiko-himicheskogo obsjestva* (Journal of Russian Physico-Chemical Society), Part : Physics, vol. 14, issue 9 a (1882), 518-535.

in German science. On these pragmatic and positivist grounds, modern science seemed to be accepted by the Russian culture.

As early as the middle of the 18th century, physicists were concerned with metaphysical problems. For example, these problems arose as the principle of least action was taken under discussion in the 1750s. There was an exchange between the Cartesian party (P. Maupertuis, L. Euler) and H. Wolff, the famous follower of Leibniz, in the course of debate on this principle.

It is worth mentioning that the Saint-Petersburg academicians Euler and Daniel Bernoulli adhered to Cartesian mechanical science and despised H. Wolff's metaphysics. Their colleagues Richman and Lomonosov seemed to share this view (although Lomonosov studied physics under Wolff). As H. Wolff had been suggested by Leibniz and proposed by Russian authorities to take the position of President of the Saint Petersburg Academy of Sciences and was engaged to nominate candidates for the position of Saint Petersburg academician, debates between him and the Cartesian party was in touch with the Saint Petersburg Academy.

Yet at the beginning of the 19th century, physicists did not usually care if metaphysical problems were settled. This non-metaphysical physics were accepted by the secularized Russian culture.

THE CULTURAL BACKGROUND

Why was the Russian culture so late in giving an adequate response to the Scientific Revolution (or revolutions) connected with Galileo, Descartes, Huygens and Newton ? This question entails taking into consideration some characteristic features of the Russian intellectual culture of the corresponding periods and eventually culminates in the issue about the Russian Orthodox Church's attitude to science. The point is that Russia in the 16th and 17th centuries did not have a rich secular culture[8]. In fact, in spite of an active secularization, the Russian culture of the 18th century was religious in its educational backgrounds[9].

As was noted above, the Russian Orthodox Church's attitude to modern science was favourable. In his fundamental book, *Essays in the History of the Heliocentric World View in Russia,* the prominent historian of science B.F. Raikov aimed to describe how the theory of Copernicus was accepted by Russian science, education and culture[10]. As a loyal Soviet historian Raikov tended to reconstruct the context of the struggle between science and religion.

8. P. Bushkovich, *Religions and Society in Russia. The 16th and 17th centuries,* Oxford Univ. Press, 1992, 4

9. G. Marker, " Faith and Secularity in Eighteenth-Century Russian Literacy, 1700-1753 ", *Christianity and the Eastern Slaves,* Univ. of Calif. Press, 1994.

10. Moscow, Leningrad, ed. AN SSSR, 1937.

However, he could not find any example of a strong attack again Copernicus in Russia. He found some exemptions by observing the reign of Elizabeth (1750s). He pointed to the proposition of the Saint Synod to suppress B. Fontenelle's book *Talks about plurality of the Worlds* translated into Russian by Antioh Kantemir. But Raikov himself noted that this was a delayed response to the book published 15 years before and bought up. The second exemption was a censure on *Spiritual Verses* written by Alexander Petrovich Sumarokov. This censure resulted from information given to the Synod by another Russian Poet, Kirill Trediakovsky, who envied Sumarokov and pointed to the " immoralism " of Sumarokov's verses. Sumarokov's " Spiritual Verses " did contain the idea of the plurality of the Worlds, but this very popular idea was admitted by Tredikovsky as well.

It is natural to conclude that the controversy Heliocentrism-Geocentrism was considered by the Russian Orthodox Church to be a secondary, auxiliary problem, which was not worth careful theological (theoretical) discussion.

Epiphanii Slavinetskii, who was mentioned above as a rigid orthodox theologian and leader of the anti-Latin party in Moscow, translated one of the Western *Cosmographias*, in which the theory of Copernicus was presented, and disseminated the ideas of new astronomy[11].

This tolerance toward science may be traced back to a Russian religious book of the 16th century. In the 1540s under the guidance of Moscow Metropolitan Makariy, the so-called *Velikie Chet'i Minei (Great Chet'i Minei)* was issued. That was a collection in 12 volumes, more than 14.000 pages, containing teaching, hagiographical, exegetic writings " recommended for reading ". To produce this great book a lot of work on manuscript research and many translations had been done. Writings included into this book were ordered by days of a calendar year, every day being devoted to a Holy man (Sanctus). In the space of this day everything that was connected with this Holy man was printed : his biography (Zitie), works *in extenso, etc*[12].

It is interesting that in this book under the corresponding days there are two Greek writings, containing the astronomo-geographical world view. On the one hand, *Cosmography* by Kosmas Indicopleustes (6th century) informs us that the Earth is a plane quadrangle surrounded by the Ocean and covered by the firmament (the sky) which is fitted to the four sides of the Earth. On the other hand, *Shestidnev* by John Bulgarian (10th century) admits the Ptolemy System of the World and contains advanced and the best astronomical and geographical information for the period (in particular the data for the dimensions of celestial bodies, the distances from them to the Earth, *etc.*).

11. V.K. Kuzakov, " Osobennosti nauki I tehniki srednevekovoi Rusi ", *Estestvenno-nauchnye predstavlenia drevnei Rusi* (Natural Sciences in Ancient Russia) (Moscow, 1978), 15

12. N.S. Trubetskoi, " Lectures on Ancient Russian Literature ", *Istoria, kultura, iazyk* (History, Culture, Language) (Moscow, ed. Progress, 1955), 544-616.

Great Chet'i Minei had been composed very carefully and tested many times. The fact that the book contains (from a scientific point of view) the rival world views, provides evidence that no one had been legitimated by the Church and to choose between them was not the Church's problem.

It is worth noting that Cap. II, 5 *De caelo* of *De fide orthodoxa* written by St. John Damascen (8th century) led to such tolerance too. St. John presented all the important astronomic-geographical world views of that time in it. As a conclusion he said that he did not intend to decide what view was correct, since every view left theology untouched.

This book was the most authoritative one on dogmatics in the Orthodox Church. The point was that, in contrast to Western scholasticism, orthodox theology did not canonize Ptolemaic astronomy.

In contrast to the Catholic and Protestant Churches infected, as the Russian philosopher A.S. Chomiakov told, by the sin of rationalism, the Russian Orthodox Church did not tend to work out any universal theological system covering all the world (as Thomas Aquinas'), and Russian theologians did not try to combine dogma and physics.

The East Orthodox Divinity had an essentially mystical character. Scientific knowledge, which is not mystical by definition, was repudiated by it as something useful, but secondary and auxiliary. In *Triades...* (struggling against the Byzantine followers of Thomas Aquinas) St. Gregory Palama, the great orthodox theologian and mystic (14th century), said that " external knowledge " (philosophy, astronomy, physics, *etc*) " ...is a good means to exercise an inner eye but to be concerned with it at one's old age is bad... " (I, 1, 6) and " ...we are not intended to prevent anybody from studying in external sciences, if he is not a monk, but we don't advise to devote all one's life to them and do not allow to expect that they will bring certain knowledge of divine entities " (I, 1, 12). This is a typical attitude of the orthodox theologian.

As a result the Church was tolerant toward modern science. However there was an other side to the coin. The Church did not take scientific theories into the context of theological discussions. Clergymen, who were most well-educated people in those centuries, took the scientific world-views as curious but secular and hence not very important.

By the beginning of the 19th century, the Russian culture had drastically changed. The expanded body of secular culture, that was able to adopt modern calculating mechanical science, had arisen. As was noted above, by these years modern science had changed too : it had become distanced from metaphysical discussions, which intruded into the field of theology.

SCIENCE IN EIGHTEENTH-CENTURY PORTUGAL
THE NATURALIST CORREIA DA SERRA

Ana CARNEIRO, Ana SIMÕES, Maria Paula DIOGO

The Portuguese botanist Correia da Serra (1750-1823) is among those intellectuals who mostly engaged in the process of dissemination of the new sciences in Portugal, during the eighteenth century. The leading contributors to this process were the *estrangeirados* (Europeanized intellectuals), an informal network[1] composed both of Portuguese who had contacts with European intellectual circles, or foreigners who established themselves in Portugal.

Especially during the second half of the eighteenth century one of the strategies followed for the implementation of a new scientific discourse relied to a great extent on legal and administrative measures, of which the reform of the University of Coimbra (1772) and the creation of the Royal Academy of Sciences of Lisbon (1779) became paradigmatic. In this context the first Portuguese men of science propagating a more scientifically-minded discourse, in which science was often endowed with a fundamental utilitarian dimension,

---

1. A concept of network was used, which highlights two dimensions : the scope of problems addressed and the density of relationships among individuals. In the case under consideration the network is neither formally institutionalised nor does it affirm a common strategy. Rather it represents a group of intellectuals belonging to different disciplinary fields simply sharing a common goal — the modernisation of the country. It is from this standpoint that the *estrangeirados* built up a dialogue and kept regular contacts among them, either personal or epistolary or, indirectly, through their written works. The network of *estrangeirados* becomes in this way a homogeneous though fluid structure emerging from the Portuguese intellectual community. See D. de Solla Price, *Little Science, Big Science*, New York, 1963 ; D. Crane, " Social Structure in a Group of Scientists : A test of the Invisible College Hypothesis ", *American Sociological Review*, 34 (1969), 335-352 ; B. Griffith, N. Mullins, " Coherent Social Groups in Scientific Change ", *Science*, 177 (1972), 959-64 ; H. Collins, " The Place of " Core-Set " in Modern Science : Social Contingency with Methodological Propriety in Science ", *History of Science*, 19 (1981), 6-19 ; D. Edge and M. Mulkay, *Astronomy transformed : the Emergence of Radio Astronomy in Britain*, New York, London, 1976. See also A. Simões, A. Carneiro, M.P. Diogo, " Constructing Knowledge : Eighteenth-century Portugal and the New Sciences ", *Archimedes*, 2 (1999), 1-40 ; and A. Carneiro, A. Simões, M.P. Diogo, " Enlightenment Science in Portugal : The *Estrangeirados* and their communication networks ", submitted for publication.

emerged. Such was the case of the *estrangeirado* and naturalist Correia da Serra.

CORREIA DA SERRA'S LIFE AND CAREER

José Francisco Correia da Serra was born in 1750, in Serpa, Southern Portugal. When he was six years old, his family moved to Italy for reasons still to be completely clarified. Young Correia da Serra spent about three years in Naples, where he was taught by Abbot Genovese. Then he moved to Rome, where he established contacts with the Portuguese pedagogue and *estrangeirado*, Luís António Verney (1713-1792), whose work harshly criticised the scholastic teaching practices of Portuguese universities. By his early twenties, he was already corresponding with the Swedish botanist Linnaeus. He also became acquainted with another *estrangeirado*, the Freemason intellectual Duke of Lafões (1719-1804), a former colleague of his father at the University of Coimbra, who greatly encouraged his intellectual pursuits. Correia da Serra was appointed Presbyter in 1775, although throughout his career he seems to have endorsed anticlerical attitudes. The intellectual trends of his time soon attracted him and he became particularly interested in natural history.

In 1777, following the death of King José I (1750-1777) and the political fall of his Prime-Minister, the Marquis of Pombal (1699-1782), Correia da Serra returned to Portugal, probably upon suggestion of the Duke of Lafões. In Lisbon, Correia da Serra was to collaborate with him in the foundation of the Royal Academy of Sciences of Lisbon (1779). Despite the age-gap between Lafões and Correia da Serra, the former saw in his *protégé* an enthusiast collaborator in the dissemination of new ideas pertaining to encyclopaedism and science.

In 1786 for unknown reasons, Correia da Serra sought refuge in France. He came back to Portugal two years later and was appointed Secretary to the Royal Academy of Sciences[2]. In this capacity he read an eulogy to his fellow American Freemason Benjamin Franklin on 4 July 1791, shortly after the establishment of diplomatic relations between Portugal and the USA. Franklin had been elected as corresponding member of the Academy of Sciences of Lisbon in 1782, and although Correia da Serra never met him personally he highly regarded Franklin's contributions to science.

In 1795, Correia da Serra was once again forced to leave Portugal, this time for England, because he had given protection to a French Girondist liberal refugee, the physician Broussonet, who was even kept in hide at the premises of the Royal Academy of Sciences. He was persecuted by Pina Manique (1733-1805), Superintendent of Police of Queen Maria I, who for a long time was

---

2. He succeeded the Viscount of Barbacena who had meanwhile been appointed to the position of Governor of Minas Gerais, one of the richest Brazilian regions in gold and other minerals.

suspicious of the activities of both Correia da Serra and of the Duke of Lafões and, generally, of all the members of the Academy of Sciences. Pina Manique considered the interest of that institution in French authors as an implicit endorsement of the liberal cause, and consequently as a subversive threat. In fact, Abbot Correia da Serra belonged, as many other Portuguese intellectuals, to Freemasonry, an organisation which, in Portugal, was deeply linked to the bases of a new political, ideological and intellectual framework epitomised by liberalism.

In England, Correia da Serra had the possibility of developing close friend-ships with James Edward Smith, whom he held in great esteem[3], Robert Brown[4], Richard Anthony Salisbury and Sir Joseph Banks. During this period he published various papers in the *Philosophical Transactions of the Royal Society,* of which he was a member, and in the *Transactions of the Linnean Society*, of which he was also an associate.

During six years Correia da Serra held the position of adviser to the Portu-guese Diplomatic Legation in London, probably thanks to the influence of the Duke of Lafões, uncle of the Queen. However, due to disagreements with the Portuguese ambassador the gloomy prospect of exile fell on him again. Once again the Portuguese botanist was entangled in a plot which forced him to seek refuge this time in France, where he lived from 1802 until 1812. In this country he built up a well-established reputation as a botanist and became scientifically and friendly linked to Lafayette, Alexander von Humboldt, Augustin-Pyramus de Candolle, Cuvier, Dupont de Nemours, Antoine Laurent de Jussieu, Julien de la Mettrie and others.

After his stay in France, Correia da Serra did not return to Portugal, but he left for the USA, instead. Thanks to the reputation he had built up in France and England as a naturalist, in 1812 he began teaching various courses on bot-any, in Philadelphia.

In America, the Portuguese botanist consolidated further his reputation of an *avant-garde* intellectual. He became a member of the Wistar Party, a selected intellectual circle led by the physician Caspar Wistar, mainly composed of members of the American Philosophical Society. Meetings were held weekly, in which various subjects ranging from literature to science and philosophy were debated. Correia da Serra engaged in various scientific travels, during which he collected data for his future work in botany and geology. Throughout 1815, he lectured various courses on botany at the University of Pennsylvania,

3. Correia da Serra referred to James Smith as " (…) of all England you are the person with whom I like the better to converse and Botany is now the purest spring of pleasure to me ". Letter from Correia da Serra to James Smith dated 4 July 1795. The Smith papers, London, The Linnean Society.

4. It was Correia da Serra who introduced Robert Brown to Joseph Banks.

which were considered extremely innovative. In particular, he introduced Jussieu's system in Philadelphia and in all probability in America.

Correia da Serra paid a visit to President Madison in Washington and, in July 1813, he met for the first time Thomas Jefferson at Monticello, Virginia, beginning in this way a lasting friendship. He became such a regular visitor to Monticello that one of the guest rooms was named after him the " Abbé's Room "[5]. Later, at the request of his friend, Correia da Serra was invited to express his opinion about the Statutes of the University of Philadelphia. His remarks were highly praised by Jefferson.

Correia da Serra returned to Portugal in 1821, following the victory of the Liberals (1820). At long last, he found himself praised and honoured by his fellow countrymen apparently more by his political posture than by his scientific contributions, and was compensated by the new-born Liberal regime.

Correia da Serra had not much chance to further botanical research after his return to Portugal as he died in 1823, in Caldas da Rainha, a town not far from Lisbon.

THE EVOLUTION OF CORREIA DA SERRA'S SCIENTIFIC THOUGHT

From the methodological and theoretical point of view, Correia da Serra's main influences came from the German botanist Joseph Gärtner and from the French Antoine-Laurent Jussieu. Gärtner's work went almost unnoticed in Germany but it influenced British naturalists and in France it was highly regarded by Jussieu. Both men shared friends with Correia da Serra, in particular Cuvier, J.P.F. Deleuze, R.L. Desfontaines and Joseph Banks. The latter had been, on different occasions, patron of both Gärtner and Correia da Serra and it was surely through him that the Portuguese botanist became acquainted with the work of the German carpologist.

On the other hand, A.L. Jussieu, whom Correia da Serra met during his first exile in France, followed closely the lines of Adanson's work which developed in the period when it became clear to botanists that the Linnean artificial method of classification, i.e., downward classification based on logical division was unsatisfactory. Adanson rejected the single-character divisional method and proposed an upward classification by empirical grouping to which he called natural method. Correia da Serra adopted this alternative approach through Jussieu, but he was to introduce his own refinements, which were to influence decisively A.P. de Candolle.

5. The room was recently refurbished with a grant from the Luso-American Foundation in Lisbon. After Jefferson's death, the estate was bought by one great great-son of Diego Nunez Ribeiro, the former Portuguese physician of King João V, and uncle of other well-known Freemason and Jewish physician, the *estrangeirado* António Nunes Ribeiro Sanches, whose educational and medical ideas inspired the reforms of the Marquis of Pombal, and notably the creation of the College of Nobles (Lisbon) and the reform of medical studies at the University of Coimbra.

Correia da Serra's first article focused on the fructification of submersed algae and was published, in 1796, in the *Philosophical Transactions*. Already in this paper he shows a very peculiar way of making his points by revising critically and carefully those made by his fellow botanists in prior investigations. The question addressed by Correia da Serra was that of the reproduction of plants, especially those whose organisation though simpler was still poorly understood.

Since his object of study was submersed algae he pointed out not only the importance of understanding the relations between structure and function but also the influence of the specific environment of these plants in inducing particular modifications in their reproductive organs, stating that proper observations led him to think that algae were hermaphrodite flowers.

From this paper, the principles which were to orient Correia da Serra's subsequent research are already implicit. He assumed the unity of plan underlying Nature and, consequently, botanical classification should be based on the establishment of analogies derived from the comparison of plant structures. These structures are analysed in relation to the functions they perform, and the modifications of organs which botanists observe in plants are, in his view, induced by the environment in which they live.

The next paper written by Correia da Serra was " On two genera of Plants belonging to the Natural family of the Aurantia ", published also in the *Transactions of the Linnean Society* in 1799. The object of the paper is to examine the general characters and the natural affinities of the *Crateva Marmelos* of Linnaeus, and of the *Crateva Balangas* of Koenig which he had observed in the herbarium of Sir Joseph Banks. The Portuguese botanist considered these plants, each of which was, in his opinion in a genus by itself, not only different from the *Crateva*, but also belonging to a distinct order. Before proceeding to the demonstration of his claims, he states in more assertive terms his emphasis on affinities rather than on distinctions as a means of classification. Therefore he was conferring a supremacy to the natural systems of classification over the artificial ones, distancing himself from the Linnean tradition.

In 1805, during his exile in France he published a paper entitled *Observations sur la famille des orangers et sur les limites qui la circonscrivent* which is perhaps best known to historians because it was in this paper that the Portuguese botanist introduced the concept of " symmetry " which Augustin Pyramus de Candolle[6] used in his book *Théorie élémentaire de la botanique* (Paris, 1813). In this article he makes the crucial observation that botanists had previously been more interested in differences than in similarities. This change in focus, which at the time was beginning to develop in some quarters of natural

---

6. See A. Arber, *The Natural Philosophy of Plant Form*, Cambridge, 1950 and P.F. Stevens, " Haüy and A.-P. Candolle : Crystallography, Botanical Systematics, and Comparative Morphology, 1780-1840 ", *Journal of History of Biology*, 17 (1984), 49-82.

history, played a major role in overturning the criteria of classification of both animals and, subsequently, of plants. The idea of classifying upward (or compositional classification) which developed in this period was a major methodological revolution, Correia da Serra emerging, in this way, as a leading contributor to the establishment of this new methodology. However, the fact of the matter is that before 1805, Correia da Serra was already following these guidelines as the analysis of his prior work has revealed. The novelty of this study of the *Citrus* family, which the author uses as a model to expound his theoretical views is, in fact, the concept of symmetry, which is mentioned for the first time in his published work, but emerges as a natural outcome of his prior research. For the Portuguese botanist, symmetry was the particular arrangement of parts resulting from their respective position and form, that is, it refers to the similarities common to all members of a natural family.

Taking into account the methodological orientation of Correia da Serra's prior work and also the fact that he is at this time in contact with such eminent anatomists, such as Cuvier and Etienne Geoffroy de Saint-Hillaire, he expressed the wish that comparative anatomy should, in effect, be applied to botany. However, due to the tribulations which affected his life he had been unable to pursue this line of research. Within his studies on carpology, a newly created branch of botany whose foundations had been laid down by Gärtner, Correia da Serra states more precisely his purpose of establishing the relationships of plants' internal organisation. He distinguishes this approach from the traditional one which aimed at describing and classifying plants by analysing their external structure.

Although Correia da Serra does not mention the word symmetry in the *Observations carpologiques*, this notion is implicitly present since his purpose is to find the basic plan of a group, organ system or organ and ascertain its variations and their significance. Both in this article and in the one which follows, *Vues carpologiques*, it is particularly clear that in Correia da Serra's botanical work plants acquired a third dimension in depth since in his view botany could no longer remain at the level of the analysis of visible structures. Despite being different from animals so that the analogies between both have to be carefully circumscribed as he had shown in previous articles, plants share with animals the fact that they are organised beings. Consequently, only when a similar tool, comparative anatomy, is applied to plants can their structural and functional regularities taken as variations of a same plan be unveiled.

The link which should be made between structure and function is once again emphasised and the author entreats botanists to focus on plant physiology which, arguably, had been ignored by his predecessors and contemporaries. Such an approach, at the time being greatly developed in zoology, places Correia da Serra in the forefront of early nineteenth-century botanical studies and makes him a leading advocate of the application of this methodology to botany.

CONCLUDING REMARKS

The present analysis of Correia da Serra's life and scientific work shows that, despite his relatively small scientific production, he became a botanist whose contributions were acknowledged and recognised by his most eminent European peers. He played an important role by promoting scientific communication between natural historians of different nationalities, contributing in this way to an effective exchange of scientific knowledge not only between the centre and the periphery of Europe, but also within the centres and the peripheries. Living in a period of Portuguese history marked by deep political changes he endured, as most of his fellow *estrangeirados*, the direct interference of state politics in the shaping of his life and career.

At the scientific level, Correia da Serra's contributions were extremely innovative when put in the European context of his time. Correia da Serra's work had not much impact upon Portuguese botany, but his ideas constituted the seeds on which foreign botanists developed and built up their own. In effect, the Portuguese botanist was unable to produce a comprehensive work such as a treatise on botany in which he could have expounded his methodology in a deeper and more comprehensive manner. The reasons may have been many : the fact that he came from a peripheral country ; the unrest of his life marked by frequent exiles, and the circumstance that he often had to depend on the good will of his friends and colleagues, all may have contributed decisively to prevent him from giving to his botanical work a greater projection.

His work as a botanist was carried out in the realm of taxonomy and systematic classification, an area which was not only at the origins of Botany, but was also one of its bases. Abbot Correia da Serra favoured an evaluation of all characteristics of plants as the starting-point for their identification. He focused on the dissection and description of fruits and seeds of more than 20 species, mainly belonging to the family of orange-trees, basing his work essentially on similarity criteria.

Underlying the botanical investigations of Correia da Serra were transformist arguments, and in general terms, two different ways of conceiving nature and the sciences of life. On the one hand, the systems of classification constructed from a creationist and fixist structure had as a purpose the individual classification of a specimen. On the other, the possibility of establishing the mechanism which allows to proceed from one species to another, implies the assumption of diachronic transformations in which actual species developed from ancient forms, i.e., implies a transformist thinking. At the very heart of this new vision of the living world were the notions of affinity, structure and function. The emphasis put on the interaction organism-environment justifies the capacity of variation and adaptation throughout time. Correia da Serra's emphasis on the study of the organs of fructification, by considering them central to the definition of degrees of affinity, shows his endorsement of this new functionalist view of the living world. When the internal organs are approached as interactive parts within the organism, their relative transformations can be grasped.

Through the establishment of affinities, the investigations on reproduction organs of plants were to allow the establishment of both synchronic and diachronic relations between species, and for this reason Correia da Serra is, from the epistemological point of view, a transition figure separating eighteenth-from nineteenth-century natural science.

Scientific Works :

- " On the fructification of the submersed Alge ", *Philosophical Transactions* (1796), 494-505.
- " On a submarine forest on the coast of England ", *Philosophical Transactions* (1799), 145-155.
- " On two genera of plants belonging to the natural family of the Aurantia ", *Transactions of the Linnean Society*, 5 (1799), 218-226.
- " On the Doryantha, a new genus of plants from New Holland next akin to the Agave ", *Transactions of the Linnean Society*, 6 (1800), 211-213.
- " Observations sur la famille des orangers et sur les limites qui la circonscrivent ", *Annales du Muséum d'Histoire Naturelle*, 6 (1805), 376-386.
- " Mémoire sur la germination du nelumbo ", *Annales du Museum d'Histoire Naturelle*, 13, 174.
- " Vues Carpologiques "/" Observations Carpologiques ", *Annales du Museum d'Histoire Naturelle*, 8, 9, 10.
- " Mémoire sur la valeur du périsperme, considéré comme caractère d'affinité des plantes ", *Bulletin de la Societé Philomatique*, 11, 350.
- " Sur l'agriculture des arabes en Espagne ", *Archives Littéraires de l'Europe*, 2, 239-404.
- " Observations and conjectures on the formation and nature of the soil of Kentucky ", *Transactions of the American Philosophical Society* (1811).
- *Reduction of all genera of plants contained in the Catalogus plantarum Americae Septentrionalis, of the late Dr. Muhlenberg, to the natural families of Mr. de Jussieu's system. For the use of the gentleman who attented the course of elementary and philosophical botany in Philadelphia in 1815*, Philadelphia, 1815
- " Notice respecting several vegetables existent in North America ", *Transactions of the Royal Horticultural Society*, 7.
- " Le Percement de l'Isthme du Suez, réussira-t-il ? ", *Revue Orientale et Américaine*, 1 (1859), 316-325.

Non scientific works :

- " Sur l'état des sciences et des lettres parmi les Portugais pendant la seconde moitié du siècle dernier ", in *Archives Littéraires de l'Europe* (1804).
- " Discurso histórico recitado na Academia Real das Sciencias de Lisboa (24 June1822) ", *Memorias da Academia*, 7 (1823), IV-XIV e o " Discurso Preliminar ", *Memorias Economicas da Academia* (1789).
- " General Considerations on the past and future State of Europe ", *The American Review*, 4 (1812), 354.
- " Sur les vrais successeurs des Templiers et leur état actuel ", *Archives Littéraires de l'Europe*, 7, 273.

Correia da Serra

Scientific Periodicals, Scientific Communities and Science Dissemination in a Peripheral Community

Antonio E. Ten

A first inventory of the technological and scientific periodicals published in Spain during the 19th century provides 522 journal titles[1], to which we must add some 400 publications on Medicine and other associated sciences[2]. In Spain, during the 19th century almost one thousand periodicals were published for the so-called Spanish " scientific community " of the time.

This datum is especially meaningful when we consider that such a scientific community was in a " peripheral " country in the European scientific scene of the century. This is even more significant if we take into account the well-known catalogue by Bolton[3], a source widely used in spite of its shortages. The catalogue compiles some 7.500 19th century publications from all over the world. From the comparison of both inventories, we can deduct that Bolton collects only a small part of the existing publications, from the most known and spread to the less known, although no rules can be set in this regard.

If, in terms of the internal history, the information contained in the less known publications is not extremely " surprising ", the fact gains relevancy from the perspective of external history and science sociology. The numerical magnitude of the Spanish data is so great as an example of peripheral scientific community, that its documentary analysis permits to analyse the scientific journalism phenomenon in 19th century Europe.

In the next pages, through the utilisation of the database on Spanish scientific journalism, we will prove its usefulness for the sociological analysis of the

1. A.E. Ten, M.C. Aragón, *Catálogo de las revistas científicas y técnicas publicadas en España durante el siglo XIX*, Valencia, IEDHC, 1996.

2. J.M. López Piñero, M.L. Terrada, *Bibliografía médica hispánica, 1475-1950*. Volúmen VIII : *Revistas*, 1736-1950, Valencia, IEDHC, 1990.

3. H.C. Bolton, *A catalogue of scientific and technical periodicals, 1665-1895*, Washington, Smithsonian Institution, 1897 (New York, Johnson Reprint, 1965).

production, the spreading and the organisation of science in a so-considered " peripheral " scientific community.

SCIENTIFIC COMMUNITIES, MONOGRAPHIC PUBLICATIONS AND PERIODICALS

A first reflection on the topic relates to the information that the analysis of the monographic publications and periodicals provide on the communities to which they are directed. In this context, it can be useful to follow the analysis of the scientific community concept given by Thomas S. Kuhn in his book on the structure of scientific revolutions and in other projects[4].

It is well-known that in the famous " Postscript " to the book, written in 1969 and published in subsequent issues, Kuhn proposes two definitions for " scientific community ", a strict one : " a scientific community consists of men that share a paradigm ", and a loose one : " a scientific community is formed by practitioners of a scientific speciality ".

The first definition is bound to the central concept of the Kuhnian construction and has given way to countless articles in relation with the definition of the term " paradigm ". The second one, also with abundant literature and not so linked to the kernel of Kuhn's thesis, is sociologically richer. Some of its most meaningful aspects are analysed.

Mainly concerned with the construction of his " normal science " model, Kuhn does not consider the differences between the 19th century and the previous ones or the 20th century, as to the information transmission process. However, any specialised publication permits to prove the importance of the analysis of these processes in distinguishing the existing situation in the different historical periods[5]. The origins of scientific journalism can be found in the 17th century and it consolidated — as a vehicle for the spreading of scientific ideas — in the 18th century, reaching its maturity in the 19th century. But until the end of this century, it did not become as important a spreading vehicle as " books " or monographic publications did in previous centuries. Even though the situation differed in the different disciplines and knowledge areas, scientific articles were not relevant until the 20th century as independent and self-sufficient literature able to support the dissemination of a theory. We only need to read the literature recommended by any university degree at present and compare it to that of the last century. Even nowadays, in the first university cycles, monographic publications are still a fundamental vehicle for the spreading of normal science paradigms.

4. T.S. Kuhn, *The structure of scientific revolutions*, Chicago : Chicago Univ. Press, 1962 ; *The essential tension*, Chicago, Chicago Univ. Press, 1977, esp. cap. XII.

5. B. Houghton, *Scientific periodicals. Their historical developement, characteristics and control*, London, Clive Bingley, 1975.

If we accepted this statement, we should conclude that, until the end of the 19th century, scientific journalism was secondary — in relation with books — in the characterisation of the information process of scientific communities, in the first sense of the term used by Kuhn. The situation seems contradictory, however, if we take into account Kuhn's second scientific community concept.

Indeed, if we analyse in detail the conceptual wealth in Kuhn's second definition, we find the " level " concept applied to the internal structure of scientific communities : At a first level, a " scientific community " would include all " scientists ". A first level scientific community would consist of all those human beings with a " scientific interest ".

There is also a second level in which the " scientific community " is already associated to " groups of professional scientists ". According to Kuhn, for such groups the " community " is easily established and permits to define some common characteristics with respect to other communities. These characteristics would be :

1. To follow highly qualified individuals, recognised as guides.
2. To be members to professional societies.
3. To be readers of specialised periodicals.

Kuhn introduces a third level, formed by sub-groups of second level groups and defined through the same parameters. Finally, a fourth level would be constituted by the communities to which the reflections on Kuhn's paradigm concept would actually be applied.

We should not be too concerned with the problems of the fourth scientific community level, as they have been studied in detail by Kuhn's commentators. According to him, this type of community is described as a unit producing scientific knowledge. The fourth level has been the object of numerous analyses, and it is not really in line with the objective of our study. However, the analysis of the " less internalist " levels 1, 2 and 3 is directly linked to the main topic in this paper.

More specifically, from the three characteristics that define scientific communities, the most interesting one to us is number 3 : the feature which defines a scientific community is the fact that its members are periodicals readers.

Undoubtedly, Kuhn's interest in the issue stems from the consideration of the publications as processes for paradigm transmission that would select their followers depending on the publications chosen. But another much more interesting piece of information can prove the existence of such publications : unlike a book (individual, singular and independent work of a creative mind), a periodical requires the work of devoted writers and stanch readers, at least during the life of the journal. If the reading of specialised periodicals permits to define a scientific community, the existence of publications of this type is a clear sign of the fact that there are sufficiently interested readers to maintain the publication during its period of existence. This approach to the study of periodicals in a country makes the inventories of the publications a sociological analysis tool for a given society or country, and opens new avenues.

Thus, an exhaustive inventory of scientific periodicals, regardless of their importance from an internal point of view, can provide the means for identifying emerging disciplines, groups of cultivators, places of installation and even their vitality and context. The 19th century is a seminal era for many disciplines. Many of these seeds that in a given society or country fructify easily, in others do not get to be implanted or grow weak and sick. Also, this century is vital in the evolution of the social interest in science the technology. Until the 18th century, scientific communities were necessarily small. The intellectual world was elitist and access was limited by rigid social and economic barriers. With the social rise of the middle class and the progressive implementation of the right to education and culture, the population base of potential cultivators and science-interested people widened. Furthermore, with the consequences of the first industrial revolution, technology was integrated in society and the industrial activity generated groups of people with specific interests in different branches of the production. As we will see in the case of Spain, all these circumstances left tracks in those periodicals they gave way to.

THE SPECIALISATION OF PERIODICALS

A second reflection is needed when envisaging the panorama of 19th century scientific periodicals. The scientific journal gender progressively consolidated along the century. If — in the most established scientific disciplines, such as mathematical sciences — the model of a scientific " article " was already defined, in other areas consolidated models did not exist yet. In the second chapter of his well-known book on the history of scientific and technical journals, Kronick[6] tried to define periodicals and the distinction between " periodical " and " newspaper ", and compiled different approaches to the problem. Their common characteristic, rather than their periodicity or their physical traits, is that the target readers of a " periodical " have a clear thematic profile. This is the profile that the direct study of the periodical evidences.

The rest of the book by Kronick is devoted to identifying the possible types of scientific journals. Kronick splits them into original and derivative and, in a most refined classification, in substantive, proceedings, collections, dissertations, abstract, review, almanac and other. The book provides detailed information, and it is important to indicate that such types necessarily meet the needs of the types of target readers to whom they are directed. For example, original articles meet the demand of a paradigm-creating community, while small news and information collections are addressed to the general public. Again, the

---

6. D.A. Kronick, *A History of Scientific and Technical Periodicals. The Origins and Development of the Scientific and Technological Press*, New York, The Scarecrow Press Inc., 1962.

existence and the contents of the newspapers provide direct information on the scientific communities of a country and their range of interests.

However, the doubtless value of Kronick's classification clashes with the great diversity of newspapers that are published all along the century. Indeed, if the newspaper concept itself evolves during the 19th century, the " scientific " and " technological " concepts undergo an even more intensive evolution. From the beginning of the century, everyday life becomes more and more influenced by science. The scientific and industrial revolution creates society models for those who think " science " is more than a scientific theory. Science was to become a way of analysing reality. As the century went by, more and more population layers integrated science and technology in their lives and professions. All this is reflected in the titles of the publications, as their existence evidences the vitality of the conceptions and interests behind them.

Strictly scientific journals as understood at present are made up of original articles, correspondence sections, abstracts, reviews on books and relevant news, and were first published in the 18th century. Together with them, new journal types appeared which were not in line with their typical image. By order of specialisation, professional and technological journals are the following types to consider. Along the 19th century, the non-specialised nature of trade and manufacturing progressively turned into modern industry. Farmers became industrialists who followed up the technological improvements and applied them to their crops. Cattlemen shifted from manual feeding to feeding the cattle by means of machines ; fishermen got interested in the biology of the species… New trades and techniques produced emerging scientific and technical communities whose maturity was determined by the appearance of periodicals that met their needs and demands. These groups agreed with Kuhn's intention of characterising second and third level scientific communities. Only a careful analysis of these types of publications can prove the existence of communities whose reality could not be demonstrated otherwise.

Together with these two large periodical groups, the scientific wish and the fashion at the time gave way to two further groups : periodicals for the dissemination of science among non-specialised but interested people and general publications with a " touch " of modernity (achieved by using the term " scientific "). The first group is probably the most singular and illustrates the 19th century mentality change quite remarkably. The fascination for new discoveries, theories, machines, craftsmanship… created a specialised literature eagerly read by those social classes that were more aware of the deep changes under way. This group pooled the most dynamic publishers. Pleasing themselves rather than their readers, they seemed to be missionaries propagating a new revelation ; devoted publishers promoted publications that — in most cases — did not outlive their creators and sometimes the " adventure " was not worth their investment.

The fourth publication type is the most numerous and scattered, and rather complex to analyse. We could identify it by the pompous titles and subtitles used. The adjective " scientific " or the noun " science " are linked to other nouns such as literature, arts, humanities, novelties, religion, history, politics... Their contents would definitely not be in line with our present idea of a scientific journal. However, due to their numerous readers, they became the most important " scientific modernisation " avenue in the 19th century society.

PERIPHERAL AND CENTRAL SCIENTIFIC COMMUNITIES. SCIENTIFIC PERIODICALS AS SCIENCE AND TECHNOLOGY DISSEMINATION AVENUES

The division of scientific journals in four large groups can be applied to both a scientifically advanced society and an emerging one. In a developed society (" central " countries form a scientific point of view), there are scientific communities and groups of people who are not mature enough to constitute a stable community. Therefore, within the same country, different levels of " scientific " evolution can be observed. Consequently, from a scientific and technological point of view, the " central " nature of a country is not an absolute characteristic. In certain scientific and technological areas, countries and communities can be either peripheral or central.

Similarly, the " peripheral " condition applied to a whole country is not absolute, but it is linked to specific technological and scientific fields. Even countries which could be described as " central " can sometimes be defined as peripheral. Likewise, clearly peripheral countries often witness individualities and teaching chains that configure active scientific communities. Maybe one of the aspects that best characterise a scientifically advanced society is the length of its teaching chains.

It seems timely to take this subject further before studying its effects on a national inventory. In the advanced communities, specialised periodicals, especially the first two types, are supported by stable institutions and strong teams of writers but also by the continuous activity of the collaborators. They include articles from researchers from other geographical and intellectual areas. In less developed communities where the above conditions are not met, the life of a journal is at risk. Their quality or originality levels have ups and downs that are reflected in their numbers and the balance between local articles and translations of articles from other countries. Their periodicity can hardly be maintained and in many cases they simply disappear during their first year. Studies on the contents and duration permit to follow up the intellectual health and the social and economic circumstances of the community the periodicals aim at.

Therefore, a methodology based on a careful analysis of the balance between national and international news or articles, journal translations, and the average life of periodicals in different specialisation areas would decisively contribute to determining the ways in which science is spread from central to peripheral communities.

AN EXAMPLE OF A PERIPHERAL SCIENTIFIC COMMUNITY : THE SPANISH CASE

Scientific communities in 19th century Spain

The inventory of the journals published in Spain during the 19th century — with the same methodology that would be used in any similar country — permits to prove the previously commented aspects. Thus, through the journals, it is easy to identify the different level scientific communities established by Kuhn : If we reserve the clause " scientific community of first level " for a group of people interested in the scientific and technological activity, we can apply the description " second level " community to those groups of scientists and technicians who are members to certain official institutions and societies rather than individuals belonging to a specialised professional area. This group could be broken down as follows :

- Official Academies. The Spanish Science Academy : the Royal Academy of Exact Sciences, Physical and Natural Sciences of Madrid, and the Barcelona Royal Academy of Natural Sciences and Arts.

- Universities, Special Schools and Research Centres. The most dynamic universities in Spain were the Central University of Madrid, with their own scientific publication ; the rest of the universities published their Yearbooks and Reports. Among the schools, the Special School of Mines or the Industrial Engineering School also ran some publications. The Astronomic Observatory of Madrid is an example of a research centre.

- Official Administrative Centres. In Spain, the Ministry of Public Works was especially fruitful as far as the publishing of periodicals is concerned ; they published a specialised scientific journal on general topics and several specialised periodicals on relevant technical areas with information on different periods of the century.

- " Ateneos ". The Ateneos in Spain are organisations which reflect the cultural interests of a society. The publications of the Ateneo of Madrid stood out from the rest (Majorca, Palencia, Tarragona and Vitoria).

After the second level scientific communities, a third level is constituted by specialised professional groups. From the analysis of their periodicals we can derive two types of organisations : the official ones, that represent structures belonging to the administration and publish journals with their institutional support, and the private ones, with specific professional groups.

Some examples of the first type are the General Direction of Mines, with its Mine Annals ; the Hydrography Direction, with its Hydrography Annals, or the Commission of the Geological Map of Spain, with a Bulletin under the same name. The second type is probably more dynamic and significant as far as the

scientific level of the country is concerned. Through the study of their periodicals, we know of the existence and vitality of professional and scientific
groups in Spain such as :

Architects
Farmers and cattlemen
        Grape growers
                Oenologists
        Olive growers
        Grain growers
Fishermen
Railway officers
Ship officers
Engineers
        Civil engineers
        Industrial engineers
        Mount engineers
        Mining engineers
Topographers
Veterinarians
Merchants
Electricians
Telegraphers
Metal workers
Photographers
Science teachers
Metrologists

If the presence of particular scientific communities in Spain is significant,
so is the absence of other community types. In a table in his book, Bolton
includes entries related to food technology, such as Bread, Beer, or Sugar. Such
communities are not detected in Spain in a journal inventory. Similarly, we
cannot find communities related to industrial processes like gas, ceramics...
All these data open new avenues for the analysis of the structure of the Spanish
society and its relationship with the neighbouring communities.

Furthermore, each of the groups with a presence in Spain seems to support
one or many journals with different existence periods that show the extent of
the implantation of the scientific or professional area which they represent. The
inventory and its documentary treatment therefore provide not only qualitative
aspects of the structure of the involved scientific communities, but also the
details of their life through the century.

After these analyses, the next step would be the study of the journals themselves. The internal analysis of the more specialised publications contents
would indicate, on the one hand, the ascription of the scientific groups to certain paradigms, i.e. the fourth level communities described by Kuhn, capable
of defending such paradigms in Spain. A paradigmatic case would be found in
the reception of Darwinism by the Spanish population. On the other hand, in
the most general journals, the internal analysis would provide, for example, the

framework of the technological and scientific information accessible to the Spanish society at each moment and the process through which the information is received.

The Spanish scientific periodicals and their specialisation

According to the former general analysis and applying it to the Spanish case, the periodicals of a technical and scientific nature can be divided into four large groups. From the inventory, the following figures can be drawn :

- Strictly scientific journals : 72
- Professional journals : 145
- Specialised science journals : 57
- General journals with a scientific interest : 249

If the first and the second types illustrate the vitality of second and third level communities, the third and fourth categories prove the existence of a first level Spanish community. Indeed, 57 scientific journals in a century can be surprising to those who speak of the Spanish scientific lag in this century. From this datum, we can deduce that although the 19th century Spanish scientific community did not reach the sufficient critical entity to produce first-order figures, there were numerous people interested in the scientific and technological progress of the century. The fact that not a large number of internationally recognised scientists stood out from those people is one of the open problems pointed out by the analysis.

Similarly, the figure of 249 titles indicate a scientific interest, which corroborates the previous consideration, even though it refers to collectives who were less involved in the Spanish first level scientific community. From the analysis of the duration of these titles, their geographical location, and the association of scientific key words with terms related to social collectives, important lessons can be learnt. For instance, the linking of " science " or " scientific " to " clergy ", or the use of the words " families " or " Miss " in the title of a journal is a good sign of the connection of the 19th century idea of " science " by a part of the collectives to those which — without such a tool — access would be difficult.

Science reception from central countries. Periodicals in a peripheral community

As noted previously, in a more internal analysis, scientific journals provide the key to the analysis of the influences that " central countries or communities " have on more " peripheral " countries or communities. If in the second and third level communities this fact is evidenced by the number of quotations by foreign scientists included in original articles, in first level communities this is illustrated by titles which are literal translations of singular foreign journals or extracts from other publications. The French *Nature*, for example, has an almost literal translation in the Spanish *La Naturaleza*. Examples abound.

Regarding journals including summaries from the foreign specialised press, there are also numerous examples which can illustrate the power of the method. The title *Abeja*, for example, acknowledges its dependence on the German scientific culture, La *America científica e industrial* is the Spanish translation of *Scientific American*, and *El mundo científico* clearly accepts its dependence on the scientific progress and news, and French techniques. The internal analysis and systematic study of the compiled journals would provide valuable avenues for the knowledge of the transmission and of spreading of scientific ideas in a receiving community.

Finally, the tool represented by the bibliometric and documentary study of the published technical and scientific journal in a country is adequately complemented with the analysis of quotations by foreign scientists and the inventory of complete translations of original articles written in other languages. In 19th century Spain, the use of languages other than Spanish for scientific purposes was limited. The way to overcome the language barrier seems to be sufficiently known in the case of books, but as far as journals are concerned, this is practically unknown. We hope that, through the proliferation of national scientific journal inventories especially in peripheral countries in relation with the more developed nations from a scientific and technological point of view, and with the application and improvement of the methodologies indicated, the History of the Science will be able to increase its number of tools in order to understand the science transmission process in the international scientific community.

PART TWO

LATIN AMERICA

INTRODUCCIÓN
LA REVOLUCIÓN CIENTÍFICA EN AMERICA

Celina A. LÉRTORA MENDOZA

Uno de los problemas que más interés suscita en las investigaciones actuales de Historia de la Ciencia es el proceso de difusión, asunción y adaptación estandarizada de las nuevas teorias científicas. En otros términos, se trata del proceso que Kuhn ha llamado " normalización " de los nuevos paradigmes. Y si hay un paradigma " paradigmático " es sin duda la " nueva fisica " (newtoniana). En cierto sentido fue más revolucionaria — para la ciencia — que el heliocentrismo, porque tuvo implicaciones en todas las ramas del saber. Y aunque no fue tan gravemente cuestionada y perseguida como su antecesora, la teoría galileana, determinó interesantes polémicas que enriquecieron a la ciencia posterior en grados antes desconocidos.

La proyección de esta polémica fuera de la ciencia estricta ha sido el primer ejemplo histórico de situaciones antes inéditas, pero que hoy son habituales. Pensemos que durante todo el s. XVIII la discusión sobre el newtonismo motivó y originó proyectos de vastos alcances, como las expediciones geográficas o las investigaciones de química aplicada, cuyos efectos económicos, sociales y hasta políticos sobrepasaron ampliamente el marco específico de la teoría física que los posibilitó.

El caso americano no queda fuera de la polémica newtoniana. En los múltiples centros académicos de Hispanoamérica las teorías de Newton fueron conocidas y discutidas casi al mismo tiempo que en Europa. Numerosos testimonios en archivos, anticuariatos y bibliotecas aguardan estudios pormenorizados. Sin duda estas investigaciones no sólo aportarán nuevos conocimientos sobre la historia de la ciencia latinoamericana, sino que también servirán a la historia comparada de la difusión científica y contribuirán a una mejor comprensión de este tipo de procesos científicos cuya importancia prospective está fuera de toda duda.

Como es sabido la obra de Newton determinó dos líneas de continuidad científica, de caracter bastante diferente. Por una parte, *Principia Mathematica* fue el origen de una Física General de abordaje altamente matematizado (mecánica celeste, mecánica, hidráulica y teoría de las vibraciones), comprendiendo el estudio de las propiedades generales de los cuerpos (inercia, peso, impenetrabilidad, *etc.*) y que fue desarrollada en el s. XVIII gracias a los trabajos de Bernoulli, Euler, Clairaut, D'Alembert, Lagrange y Laplace. Una segunda línea de investigación tuvo como base sobre todo la *Optica* newtoniana, continuando una tradición experimentalista prenewtoniana, que en el s. XVIII en Francia tomó el nombre de Física Especial, entre cuyos cultivadores hallamos a Gravesande, Nollet, Boerhaave y Musschenbroek, a los que habría que añadir otros grandes creadores teóricos como Lavoisier y Ampère.

Todos estos autores fueron conocidos y citados por los estudiosos americanos, quienes tomaron parte activa en las polémicas teórcas propias de cada una de estas ramas de la Física. Incluso hay que destacar el hecho de que a partir de mediados del s. XVIII casi todos los centros académicos (incluso los regenteados por instituciones poco afectas al newtonismo) habían introducido el doble tratamiento de la Física (General y Especial).

Hasta hace pocos años este panorama era sólo someramente conocido y mencionado en las historias generales del pensamiento colonial americano. Pero por fortuna la situación ha variado rápidamente. Hoy podemos contar con estudios amplios y profundos del proceso en distintas subregiones de América : México, Colombia (Nueva Granada), Cuba y Argentina (Río de la Plata) con extensiones a otras zonas vecinas. En estos casos las instituciones patrocinantes de los estudios han sido sobre todo las Universidades y los Consejos de Investigación, cuyos sistemas de apoyo han permitido y facilitado el intercambio informativo y la formación de una red de investigación que permite homogeneizar métodos y evaluar comparativamente los resultados.

Con este criterio hemos encarado la organización de esta área en el Simposio sobre la Revolución Científica en la Periferia, como una pequeña pero significativa muestra del estado de la cuestón en este momento. Continuamos con una línea abierta con ocasión del III Congreso Latinoamericano de Historia de la Ciencia (México, 1992) y que se ha continuado en el IV (Cali, 1995) y en diversos encuentros internacionales. Los trabajos que ahora presentamos, aunque puntuales, muestran la existencia de un vigoroso empeño latinoamericano por conocer el pasado propio que es, con todo derecho, una parte importante del pasado universal.

Newton en México en el siglo XVIII

María de la Paz Ramos Lara - Juan José Saldaña

Introducción de las teorías mecanicistas en la Nueva España

Las teorías mecanicistas fueron introducidas en la Nueva España desde el siglo XVII, por algunos novohispanos como Fray Diego Rodríguez (1596-1668) y Carlos de Sigüenza y Góngora (1645-1700), quienes se interesaron por las teorías científicas modernas, a pesar de la censura que el Santo Oficio mantenía hacia este tipo de lecturas. Se documentaron en las obras de autores como Tycho Brahe, Copérnico, Descartes, Galileo, Kepler, Gilbert, Apiano, Cristóbal Clavio, Tartaglia, Bombelli, Cardano, Stevin y Neper entre otros ; y las difundieron en la Real y Pontificia Universidad a través de la cátedra de astrología y matemáticas, donde estudiaban tanto las matemáticas " puras " (geometría, aritmética, álgebra y trigonometría) como las " impuras ", dentro de estas últimas, se enseñaban diversos temas, entre los cuales se encontraba la mecánica, el magnetismo, la hidrostática y la astronomía. Con sus conocimientos, ambos participaron en la elaboración de propuestas para dar solución a problemas urbanos, destacando su colaboración en los problemas del desagüe de la ciudad de México, al cual no se le había encontrado aún una solución adecuada. Ambos fueron excelentes astrónomos ; entre sus investigaciones destacan la determinación de la latitud y longitud del valle de México, sus cálculos sobre eclipses, la elaboración de tablas astronómicas, la construcción de aparatos e instrumentos, etc[1]. En particular, el interés de Sigüenza por la astronomía lo condujo a establecer correspondencia con científicos europeos como Kircher, Cavina, Caramuel, Cassini, Zaragoza y Flamsteed entre otros[2].

1. La cátedra de astrología y matemáticas se empezó a impartir en la Real y Pontificia Universidad de México el 22 de febrero de 1637. Su catedrático fue el mercedario fray Diego Rodríguez. Ver E. Trabulse, *El círculo roto*, SEP 80, México, Fondo de Cultura Económica, 1982.

2. H. García Fernández, " Manifiesto filosófico sobre un genio novohispano : Don Carlos de Sigüenza y Góngora ", *Ciencia y Desarrollo*, núm. 75, año XIII (julio-agosto 1987), 41-50.

Aunque estos novohispanos son los que se han estudiado con mayor profundidad, existieron otros a quienes también les interesó la ciencia moderna, de los cuales algunos ya empiezan a ser estudiados con formalidad como son los casos de Luis Becerra Tanco (1602-1672), Fray Ignacio Muñoz, Melchor Pérez de Soto y la poetisa Sor Juana Inés de la Cruz (1648 o 1651-1695). En particular, sobre Sor Juana se sabe que realizó estudios de acústica, acostumbraba rebatir el aristotelismo, y gustaba de mostrar el movimiento de un trompo con harina en el piso para observar con mayor claridad que la trayectoria de la punta " no es un círculo sino una espiral cuya curvatura aumenta conforme se pierde el impulso "[3]. Respecto a Muñoz y Becerra Tanco también fueron profesores del curso de Astrología y Matemáticas de la Real Universidad. Este último, se presentó en 1672 como único opositor para ocupar dicha cátedra, lamentablemente murió pocos meses después de adquirirla en propiedad, ocupando su puesto el eminente Carlos de Sigüenza y Góngora[4].

Estos personajes del siglo XVII, empezaron a integrar una comunidad científica preocupada por conocer, cultivar, difundir y aplicar el conocimiento científico moderno, a pesar de las prohibiciones oficiales de las autoridades virreinales. La participación de esta comunidad fue crucial para el desarrollo que experimentó la ciencia en la Nueva España durante el siglo XVIII, periodo en el cual, se conjugaron una serie de factores que propiciaron un auge científico, considerable tanto en cantidad como en calidad que se reflejó en : la amplia difusión que la ciencia tuvo gracias a la publicación de revistas, periódicos, folletos y libros ; la introducción de la enseñanza de la física moderna en cursos de filosofía ; y la fundación de instituciones educativas con enseñanza netamente científica[5].

Algunos de estos factores fueron : el haber contado con una tradición científica que provenía del siglo XVII y cuyos antecedentes datan incluso del mismo siglo de la Conquista ; la conformación de una élite criolla, que empezaba a compartir con los españoles el dominio de la sociedad colonial ; el cambio del régimen de un gobierno eclesiástico a uno ilustrado encabezado por los Borbones españoles ; la existencia de intereses comunes entre novohispanos y peninsulares (especialmente en el campo de la minería) que hicieron viables la negociación de propuestas novohispanas ; el movimiento de Ilustración acom-

3. E. Piña, " Comentarios a la historia de la física en México ", *Boletín de la Sociedad Mexicana de Física*, 6-1 (1992), 28.

4. Becerra Tanco destacó por sus múltiples estudios y publicaciones que realizó sobre la aparición de la Virgen María de Guadalupe utilizando conocimientos de física. Por ejemplo, fundamentándose en la óptica y en la obra *Ars Magna Lucis et Umbrae* de Atanasio Kircher realizó un análisis sobre la manera que la imagen de la virgen se estampó en la tilma de Juan Diego. Ver J.M. Espinosa, *La comunidad científica novohispana ilustrada en la Real y Pontificia Universidad de México*, Tesis de Maestría en Filosofía de la Ciencia, México, Universidad Autónoma Metropolitana Iztapalapa, 1997, 24-36.

5. J.J. Saldaña, " Ciencia y felicidad pública en la ilustración americana ", *Historia Social de las Ciencias en América Latina*, México, Editorial Porrúa, 1996, 151-207.

pañado de una ideología progresista y la influencia francesa que aumentó el entusiasmo de los criollos por la ciencia moderna y la distancia y el desprecio que existía hacia los peninsulares ; la participación de una comunidad de criollos ilustrados con intereses industriales que propuso a la metrópoli una serie de reformas entre las que se encontraba la institucionalización de la física, dado que se veía como uno de los instrumentos necesarios para la modernización y del desarrollo de la sociedad[6].

La introducción, divulgación, difusión y parte de la enseñanza de la física moderna se llevó a cabo por un pequeño grupo de criollos entre los que destacan : Juan Benito Díaz de Gamarra (1745-1783), José Antonio Alzate (1737-1799), José Ignacio Bartolache (1739-1790), Joaquín Velázquez Cárdenas de León (1732-1786), Antonio de León y Gama (1735-1802), Francisco Javier Clavijero (1731-1787), Francisco Javier Alegre (1729-1788) y José Rafael Campoy (1723-1777)[7]. Para ello, consultaron no sólo los *Principia Matemática* de Newton (mismos que se introdujeron en latín) sino también libros escritos por otros autores, por enumerar algunos : *Tran. Filosoficas* de Jacquier ; *Dictionnaire Universel de Mathématique et de Physique* (1753) y *Ciencias exactas* de Saverien ; *Principios filosóficos* de Euler ; *Cours de Physique* (Paris, 1749) de Musschenbroek, *Leçons de Physique Expérimentale* (París, 1738), *Arte de Experiencias* y *Electricidad* de Nollet ; *Elementos de Física Teórica y Experimental* (Madrid, 1787), *Resumen Histórico y Experimental de los Fenómenos Eléctricos* (Madrid, 1792) de Sigaud de la Fond ; *Elémens de Physique Matématique* (Leide, 1746) de s' Gravesande ; *Traité de Dynamique* de d'Alembert ; *Cours de Physique Expérimentale* (Paris, 1751) de Desaguliers ; *Les Entretiens Physiques d'Ariste et d'Eudoxe ou Physique Nouvelle en dialogues* (1755) de Regnault[8].

Los principales medios de difusión que utilizaron fueron a través de las publicaciones (semanarios, gacetas, diarios y en general revistas de divulgación científica y técnica) y de la enseñanza de la física en colegios religiosos, en un principio, y posteriormente en laicos. El estudio de la historia de la enseñanza de la física en los colegios novohispanos nos ha permitido introducir al contenido de los libros de texto para determinar que temas de física eran los de mayor interés para los criollos, el grado de formalidad matemática que llegaron a manejar, la profundidad de conocimiento mismo y la respuesta de los jóvenes estudiantes hacia estos temas. Por otra parte, el contenido de las publicaciones nos ha abierto una puerta para determinar el dominio que alcanzaron en esta disciplina y la producción científica que emergió de la "domici-

6. M.P. Ramos Lara, *Difusión e Institucionalización de la física en México en el siglo XVIII,* México, Sociedad Mexicana de Historia de la Ciencia y de la Tecnología, Universidad Autónoma de Puebla, 1994, 29-30.

7. De recientes investigaciones han emergido los nombres de otros jesuitas como, Agustín Castro (1728-1790), Raymundo Cardán, P. J. Mariano Soldevilla y Pedro Bolado.

8. M.P. Ramos Lara, *op. cit.,* 145-159.

liación " de este conocimiento. La mayoría de estos artículos refleja un especial interés por estudiar su entorno natural, incluyendo aspectos geográficos, geológicos, astronómicos, botánicos, médicos y tecnológicos entre otros.

Sus investigaciones no sólo se limitaron a repetir experimentos o a explicar los fenómenos utilizando leyes, teorías y principios, sino también para verificar que éstas que se atribuían como " universales ", realmente lo fueran, por lo que deberían de ser principios válidos en cualquier parte del mundo, incluyendo la Nueva España que formaba parte del Nuevo Mundo, término que según algunos novohispanos, fue asignado por los europeos a un territorio cuyas características físicas eran muy diferentes a las que ellos conocían. Estas discrepancias físicas, decía Antonio de Ulloa : " les exigía a los físicos sabios y curiosos, confrontar sus principios y reglas generales y explicar las causas de las experiencias particulares que tenían lugar en el nuevo territorio "[9].

Algunos novohispanos se sintieron con la obligación de llevar a cabo esta tarea, así que en su labor de verificación surgió también la refutación a algunos de los principios científicos que no pudieron explicar fenómenos o hechos producidos en esta localidad, como veremos más adelante.

De estos estudios emergieron investigaciones y propuestas científicas originales que quedaron mucho tiempo en el olvido por no haberse difundido o dado a conocer en el Viejo Continente. Pero que sin embargo, ahora cobran importancia por ser estudiados desde un enfoque de historia social de las ciencias que incorpora elementos locales de diversa índole, como geográficos, culturales, sociales, económicos y políticos entre otros.

Las áreas de la física que se cultivaron fueron la astronomía y la física experimental. La primera de ellas, generalmente observacional, se basó en la mecánica celeste y la segunda en la mecánica desarrollada por Isaac Newton y por su escuela. A finales del siglo XVIII se introdujo también la óptica, la electricidad y el magnetismo.

LA MECÁNICA NEWTONIANA EN LA ENSEÑANZA DE LA FÍSICA

La introducción de las teorías mecanicistas newtonianas en la Nueva España tuvo lugar principalmente en los colegios religiosos. Los jesuitas fueron los primeros en insertar en sus cursos de filosofía conocimientos de física impartidos a manera de debates entre teorías escolásticas y modernas como se percibe en los tratados que algunos de ellos escribieron. Esta labor se vio obstaculizada en 1767 al ser expulsados del reino por orden del rey.

9. A. Ulloa, " Sobre las varias disposiciones que tienen los terrenos en diferentes partes de la Tierra ; y los efectos que ésta ocasiona en los temperamentos, no menos que en las producciones ", en E. Trabulse, *Historia de la ciencia en México, siglo XVIII* (Conacyt/Fondo de Cultura Económica, 1985), 29-57.

En el Colegio de San Francisco de Sales, el filipense Juan Benito Díaz de Gamarra trató de introducir como libro de texto, en la cátedra de filosofía, el que escribió con el nombre de *Elementa Recentioris Philosophiae* el cual publicó en 1774. Esta obra abordó la mayor parte de los temas considerados como parte de la física, como su utilidad, las propiedades de los cuerpos, el tiempo, el espacio, el vacío, el movimiento y sus leyes, la estática, las máquinas, la hidrostática, " la electrología " (hoy electricidad), el calor, la luz, los sistemas del mundo, la naturaleza de la Tierra, de la fuerza de la piedra magnética, la naturaleza del agua, del fuego y los meteoros, y de las plantas[10]. Al principio esta obra gozó de gran popularidad, al grado de haber sido recomendada para enseñarse en la Real y Pontificia Universidad de México. Sin embargo, el proyecto no fructificó, ya que Gamarra fue atacado en varias ocasiones y denunciado ante el Santo Oficio. Aunque éste no condenó su obra, Gamarra renunció a su puesto y su libro no se difundió más[11]. Hoy en día esta obra se ha considerado como el primer libro escrito por un novohispano en la Nueva España que estudia la física moderna[12].

En la Real y Pontificia Universidad de México, fundada en 1553, se inició la enseñanza de las teorías mecanicistas desde el siglo XVII, lo que facilitó la introducción de las tesis newtonianas a las aulas, especialmente en las clases de matemáticas, además de noticias novedosas europeas relacionadas con la física, como : noticias sobre el invento del globo aerostático y las naves atmosféricas, y listas de libros y revistas de reciente circulación en Europa[13]. Por ejemplo, la revista *Biblioteca periodica anual para utilidad de los libreros y literatos* contenía índices generales de los libros y papeles que se imprimían y publicaban en Madrid, además de indicar las librerías donde se vendían[14].

El primer establecimiento en América hispánica en el que se inició una enseñanza institucionalizada de la física fue el Real Seminario de Minería de la ciudad de México, cuya entrada en funcionamiento data de 1792. Esta institución se había venido gestando por iniciativa de los criollos novohispanos, Joaquín Velázquez de León y Lucas de Lassaga, desde la década de los setenta del siglo XVIII. La concebían como una institución para dar enseñanza técnica a jóvenes mineros y desarrollar las " artes útiles " en la sociedad. Además de

10. B. Díaz de Gamarra y Dávalos, *Elementa Recentioris Phisosophiae, Volumen Primum,* México, 1774 (Biblioteca Nacional de México).

11. V. Junco De Meyer, *Gamarra o el eclecticismo en México,* México, Fondo de Cultura Económica, 1973.

12. G. Cardozo Galué, *Michoacán en el siglo de las luces,* México, El Colegio de México, 1973.

13. *Papeles Varios,* tomo II, colección La Fragua B.N. de M. (R 626 LAF), " El globo aerostático y la nave admosférica ", Fondo Reservado de la Biblioteca Nacional de México.

14. Centro de Estudios Sobre la Universidad, Archivo de San Idelfonso, ramo : Rector, subramo : Vida académica, serie : Noticias de cátedra y sistema de enseñanza (siglo XVIII), caja 54, exp. 34, doc. 108, Biblioteca Nacional de México.

la física ahí se inició igualmente la enseñanza de las matemáticas, la química y la mineralogía[15].

La física que se enseñó y se utilizó en México en el siglo XVIII se apoyó en la mecánica newtoniana y tuvo un carácter pragmático o utilitario. En efecto, no sólo se difundió la teoría sino que se le llegó a " domesticar " y se utilizó en beneficio del desarrollo económico del país. La física experimental se utilizó para mejorar la explotación de las minas[16], y la astronomía observacional sirvió para la geografía y la geodesia. En el Real Seminario se enseñó principalmente la mecánica newtoniana. Se impartieron cursos de estática, dinámica e hidrodinámica. Las dos primeras se estudiaban para entender el funcionamiento de las máquinas simples, y la hidrodinámica para comprender el funcionamiento de máquinas más complejas como el fuelle hidráulico, la balanza hidrostática, el globo aerostático, las bombas, las máquinas neumáticas, *etc.*, como se puede observar en el libro de texto Principios de física matemática y experimental escrito en los últimos años del siglo XVIII por el catedrático de física, Francisco Antonio Bataller (1751-1800) en México.

LA MECÁNICA NEWTONIANA EN LAS PUBLICACIONES NOVOHISPANAS

La divulgación de la ciencia permitió que un grupo numeroso de la sociedad conociera y se interesara por la física y sus aplicaciones que beneficiarían a la sociedad misma. Uno de los elementos importantes fue destacar la brillante labor de los científicos, entre los que sobresalió la imagen de Newton y de quien se mencionaban frases de gran admiración como : " el celeberrimo filósofo matemático inglés "[17] ; " sublime Newton "[18] ; " el gran Newton " ; " el célebre Newton " ; " el excelso Newton " ; *etc.*

La física se convirtió en una de las disciplinas favoritas entre los novohispanos, por su carácter experimental y sistemático, que los llevó a adquirir libros, revistas e instrumentos que contribuyeran al estudio de esta disciplina. Así también iniciaron una serie de publicaciones locales donde daban a conocer, desde noticias científicas novedosas y espectaculares hasta los resultados de sus investigaciones. Algunas de las publicaciones fueron periódicas, como : *La Gaceta de México, El Diario de México,* y *El Mercurio Volante* entre otras.

En lo que se refiere a la difusión de las teorías de Newton en revistas tanto periódicas como únicas (esto es a un público más especializado), pocas veces

15. J.J. Izquierdo, *La primera casa de las ciencias en México,* México, Editorial Ciencia, 1955.

16. La minería de la plata constituía la principal actividad económica de la Nueva España. Ver D.A. Brading, *Mineros y comerciantes en el México Borbónico (1763-1810),* México, Fondo de Cultura Económica, 1975.

17. J.I. Bartolache, *Mercurio Volante* (1772-1773), Introducción de Roberto Moreno, México, Universidad Nacional Autónoma de México, 1983.

18. J.A. Alzate y Ramírez, " Observaciones sobre la Física, Historia Natural y Arte Utiles ", en edición, introducción, notas e índices por Roberto Moreno de los Arcos,*Obras* I Periódicos (México, Universidad Nacional Autónoma de México, 1980), 224-241.

se profundiza o aborda su contenido, generalmente se citan los autores o científicos considerando que ya son conocidas sus aportaciones. Así, por ejemplo,
León y Gama en su artículo sobre las auroras boreales cita a : Mussembroek,
Mairán, Maupertius, Burmann, Mayer, Tacquet, Recciolo, Grinaldo, Krafft,
Para, Euler (*Principios Filosóficos*), Cassini, de Luc (*Tratado de barómetros y
termómetros*), Cramer, Halley, Gasendo, Roemer, Kirch, Neve, Celsio, Franklin, Samuel Clarke, Newton, Mariotte, Bouguer (Traité d' Optique), Horro-
bow (Elementos de física), Hanovio, Mac-Laurin, La Lande (Astronomía),
Gravesande y Euclides[19].

El interés por la física, entre los novohispanos, no se limitó a las teorías
newtonianas, sino en general a la física moderna, que incluía, además, aspectos
de astronomía, óptica, electricidad y magnetismo entre otros campos. Las
publicaciones que llevaron a cabo alrededor de estos temas, se pueden dividir
en dos categorías :

Divulgación del conocimiento científico europeo

Las publicaciones de divulgación estuvieron dirigidas a un público no especializado, de ahí la sencillez y la poca profundidad con la que se estudiaron
algunos temas de física relacionados con la luz, la electricidad, la atmósfera, el
movimiento perpetuo y los sistemas del mundo, entre otros. En algunos de los
artículos se observa el interés de convencer a la sociedad de los beneficios que
la ciencia podía brindar. También dedican algunos de ellos a explicar el funcionamiento y construcción de máquinas e instrumentos científicos como el barómetro, el termómetro y la máquina eléctrica.

Difusión de investigaciones novohispanas originales

Las investigaciones que realizaron los novohispanos destacaron por su ori-
ginalidad y por abordar aspectos como los siguientes :
- Explicación de fenómenos naturales usando conceptos de física ;
- Uso de la física y la astronomía para explicar y resolver problemas de la
localidad ;
- Uso de instrumentos científicos como el barómetro, termómetro y telesco-
pio para hacer investigaciones propias ;
- Observaciones astronómicas e investigaciones originales.

Estas investigaciones se caracterizan por su diversidad y por el interés de
explicar su entorno natural. Sus preocupaciones principales se concentraron
en : aspectos geográficos (determinación de latitud y longitud) ; fenómenos
astronómicos (paso del cometa, tránsito de Venus por el Disco del Sol, eclipses
de Sol y de Luna, Auroras Boreales) ; fenómenos naturales terrestres (temblo-

19. A. León y Gama, " Disertación física sobre la aurora boreal ", en E. Trabulse, *op. cit.*,
1895, 238-266.

res, tipo de suelo, de vegetación, fauna, minerales, clima, rayos) resolver problemas locales urbanos como el desagüe del Valle de México ; construcción, funcionamiento y uso de instrumentos científicos útiles a la sociedad, *etc.*

La física aparece de manera implícita, al usar conceptos, teorías, principios, métodos, instrumentos, etcétera, con la intención de entender, explicar y estudiar los fenómenos antes mencionados.

En estos artículos no solamente se reproducen experimentos y se tratan de estudiar fenómenos al estilo europeo, sino verificar que los principios y teorías propuestos en Europa continuaron siendo válidos y por lo tanto " universales " al tratar de explicar cualquier hecho ocurrido en la Nueva España. En algunos casos se observa que los principios no se cumplen, así que se realizan formales verificaciones e interesantes refutaciones, que aparentemente no se conocieron en Europa.

LOCAL AND GLOBAL SCIENCE IN THE EIGHTEENTH-CENTURY HISPANIC WORLD

Antonio LAFUENTE - Juan PIMENTEL

INTRODUCTION

The aim of this paper is to assess how and to what extent local contexts re-elaborate and re-construct scientific practices and theories in the processes of globalization of science. In order to show this we have selected a number of case-studies all gathered from the eighteenth-century Hispanic world, a complex and rich colonial scenario in which modern science experienced grafts, mutations, resistance to innovation and polemics. It is not our wish to conceal our interest in avoiding current reductionist explanations : diffusionist theories, the automatic identification of science with emancipation or the radical oppositions such as creoles vs. metropolitans or ancient vs. modern hide rather than shed light on the topic studied here because they simplify excessively realities that are, actually, more dynamic, more patent, more pluralistic. On the contrary, we deem it necessary to highlight a series of phenomena that, as we historical (due to its social and human nature) view of the movements and metamorphoses development[1].

---

1. Diffusion of science, dynamics centre-periphery and matters of place and locality have become an important argument in the discipline of history of science. Some of the most influential works about this general topic discussed here are : G. Basalla, " The Spread of Western Science ", *Science,* 156 (1967), 611-622 ; L. Pyenson, *Cultural Imperialism and Exact Sciences : German Expansion Overseas 1900-1930*, New York, 1985 ; R. Mac Leod, " On Visiting the " Moving Metropolis " : Reflections on the Architecture of Imperial Science ", in N. Reingold & M. Rothenberg (eds), *Scientific Colonialism : A Cross Cultural Comparison*, Washington, 1987, 217-249 ; X. Polanco, *Naissance et développement de la science-monde*, Paris, 1989 ; D.W. Chambers, " Period and Process in Colonial and National Science ", in Reingold & Rothenberg, *op. cit.* ; D.W. Chambers, " Locality and Science : Myths of centre and periphery ", in A. Lafuente, A. Elena y M.L. Ortega (eds), *Mundialización de la ciencia y cultura nacional*, Madrid, 1993, 605-617.

From the study of such movements and changes of science in the Hispanic culture during the Enlightenment stem a number of facts that, because of their relevance, must be underlined. First, the fact that modern science did not land on waste land but on different nuclei, each with deeply-rooted local scientific traditions, on cultures in which the former settles and recreates itself. Second, modern science did not expand throughout the viceroyalties in spite of the metropolis (Spain) or against it, an argument brandished very often and whose excessive use has given rise to the demonstrative corollary that assumes that modern science and political independence are the same thing. Third, neither creoles nor metropolitans were groups as homogeneous as they are usually depicted to have been. Fourth, the criteria imposed by Western culture during the Scientific Revolution to determine whether something was modern or not, were the criteria of a triumphant culture and, as such, tend to underestimate and ignore alternative ways of understanding and living modernity.

Before dealing with specific cases, let us examine the imperial context where they sprang up. The introduction of new disciplines (astronomy, mechanics, chemistry, botany) into the viceroyalties was inextricably linked to the political programme pursued by Borbonic reformism, a programme that (not taking specificities into consideration) had as its ultimate aim to transform the old Universal Monarchy into a colonial Empire. It was an attempt to update an obsolete structure, typical of the seventeenth century and successful then, but unable to resist the pressure that transatlantic commerce put on modern international relations. Most reformist politicians did not only think of America as the main problem of the Monarchy but also as its remedy. The longed-for regeneration, to stop the decline (the topic *par excellence* of seventeenth-century Spain) involved the re-structure of American territories and the exploitation to the full of their fiscal, political and natural resources. Science, considered against this background, was employed as a suitable instrument to fulfil imperial ambitions. To set up cabinets of natural history and botanic gardens, to create astronomical observatories and seminars on mining, to organize dozens of scientific expeditions in order to chart coasts or classify plants, were all initiatives that, from the metropolitan perspective, constituted different pieces of the same puzzle.

The effective implementation of this ambitious programme entailed, none the less, the undertaking of multiple actions and tasks that surpassed and broadened its primitive identity significantly. The immediate outcome of this politicization of science to a great scale was the subsequent politicization of scientists themselves. These new actors went, in the span of a few decades, from being mere savants versed in the works of Newton or Linneus to members of an international corporation subservient only to the obligations derived from their belonging to the Republic of Letters. Many of them faced the quandary of whether confronting this new internationalist ethos within the framework of a despotic Monarchy or outside of it. Many of them, likewise, and as a result of their gradual institutionalization in the corridors of power, began to demand a greater say in decisions and bigger investments in the carrying out of their activities. As a result, conflicts between the different actors involved in this power struggle (the metropolitan power, the viceroyal one, scientists themselves) arose, with each lobby laying claim to the patriotic and utilitarian ideals of enlightened science, with each waving the banner of Enlightenment in each own way.

Thus, from the middle of the century onwards, it can be appreciated how every scientific activity, every little improvement in mining works, every hydrographic task, every astronomical observation, becomes a testing ground for experimenting the methods and *tempos* of a new imperial policy. Alongside the appearance of these new actors, new institutions, new languages and new ways in which the former interface with the ruling élites emerge. That which is now noteworthy is, we believe, not the existence of good and bad scientists, the publication of praiseworthy texts or whether the arrival of the ideas of authors such as Newton or Buffon took place sooner or later. The crucial point, beyond the degree of excellence achieved by science in the periphery, is the latter's ability to articulate other discourses on reality. To express it in a few words : its capacity to colonize the political imaginary, capacity that manifests itself everywhere, though with differences in emphasis and often with opposite meanings. The introduction of Linneus's systematics, Lavoisier's chemistry or Newton's physics, fostered with identical zeal in all the kingdoms of the monarchy, gave rise to a different set of institutions in every place, whose main actors and programmatic aims differed from each other.

COLONIAL CONTEXT

It is pointless to remind that our evaluation of the so-called colonial context cannot, nor is intended to be, exhaustive. In fact, we have decided to confine our work to two geographical locations that were both colonial centres : New Spain and New Granada. Among all the available subjects open to discussion, we will confine our comments to the argument regarding the development and repercussion of those projects undertaken under the auspices of Caldas or Zea in Bogotá and Mociño and Montaña in Mexico. This choice has been drastic but is, we think, however, an opportunity to generalize (without neglecting local particularities) and test the thesis we put forward here in the entire colonial space.

New Granada : motherland and patron

Colombian historiography has interpreted the development of Enlightenment in New Granada as an expression of continuity that will span, from the bottom upwards, the period that begins with the arrival of Mutis in Santa Fé up to the political clampdown set in motion by general Morillo, reaching its climax at the turn of the century and embodied in the works of writers like Caldas, Lozano and Zea, among others.

At about the end of the eighteenth century criticism against the hitherto almighty Mutis spreads throughout the colony, heralding a period of thorough reappraisal of the aims of the Royal Botanical Expedition, as well as of its alleged achievements. Conditions had changed radically by the time criticism against the sage Mutis was not only the preserve of university scholasticism. Thus, we see the autodidact Caldas claim without hesitation in September of 1808, after examining the writings left by Mutis in the *House of Botany*, that " now that I have managed to detect the lacunae and gaps it contains [...] I wish to salvage from the ruin that threatens the Flora of Bogotá at least my botanical writings on the Southern part of the viceroyalty... "[2].

In order to understand this change it is necessary to dwell on the scathing criticisms made by Caldas, Lozano and Zea against the way in which Mutis handled all the affairs regarding the expedition, even the building of the Bogotá Observatory itself. In this matter, his arrogance drove him to commission a building that, in imitation of the outdated models first erected in Greenwich and Paris, was already obsolete by the time construction works finished. It seems that he who had served as an adviser on so many and different matters, was not humble enough to ask for advice himself before erecting a monument to the past, rather than a useful building. And, thus, he ordered to erect a building whose inside disposition of empty spaces and observation rooms followed so closely the original European layout, that it proved completely inefficient in Bogotá, since (as everyone knows) in tropical areas heavenly bodies do not reach their highest point in the same spot as they do in the Northern hemisphere.

This case, however, does not only exemplify the negative side-effects caused by an inadequate transfer of technology, but is, as well, an episode of the struggle to undermine the power accumulated by the director in charge of the Royal Botanical Expedition. Such disputes occurred also in the scientific field ; in fact, Caldas and Pombo, Valenzuela, Lozano and Zea later, would submit their own different expeditionary project as a pragmatic alternative to the previous academic strategy adopted by Mutis. Its more or less theoretical or speculative approach was not the only subject under discussion, but, as Restrepo remarked, the public or private nature of the enterprise as well. Numerous testimonies show the difficulties experienced by Mutis when trying to establish where his vision of the expedition as a personal endeavour ended (with regard to the house of botany, the library, appointments, *etc.*) and where the scientific mis-

---

2. Cit. in S. Díaz Piedrahíta, " Caldas y la Historia Natural ", in AA. VV., *Francisco José de Caldas*, Bogotá, 1994 111-123, p. 115.

sion, financed with viceroyal or metropolitan funds, started. By the end of the 1700, a new form of patriotism which gave priority to the useful and public dimensions of knowledge begins to take shape, one that shifted priorities to different scientific disciplines, from astronomy to geodesy, from botanical systematics to the geography of plants and from exploration of territories to regional econometrics. In a few words, there were calls to substitute more organicist ideals for classificatory ones, to replace observation with mensuration, substitute the laboratory for the university chair. All this was undertaken following guidelines that would allow both to confuse and identify motherland and *patron*[3].

Their support of such ideas led them, inevitably, to criticize the Mutisian model of science. We have already referred to some of the shortcomings attributed to him. We will underline now two more, following the recommendations found in the *Proposal for the Re-organization of the Botanical Expedition* (1802) written by Zea during his exile in Europe[4]. That the expedition was to have an itinerant nature is clearly justified, in accordance with the pattern followed in those sent to New Spain and Peru, to reach a more enduring compromise with the territory and those living in it. Criticisms against the static approach adopted by Mutis when organizing his (conceived as a headquarters, the House of Botany in Mariquita, from which a widespread network of correspondents spread throughout the viceroyal domains operated) entailed a shift in those tasks to be carried out from the iconic to the observational, from the classificatory to the utilitarian, from the study to field work. The second topic, related to the demands also formulated by some Creoles from New Spain, focused on the need to take advantage of indigenous knowledge on the productive or therapeutical uses of local flora. Doubtless, the aforementioned corre-

3. Such drastic changes were implemented in a way unthinkable of a decade earlier. To illustrate this point it would suffice to quote part of Caldas's article in the *Semanario del Nuevo Reino*, of which he was editor between 1808-11 : " To observe the skies for the sake of observation would be a legitimate activity, but it would be nothing but a fruitless activity [...] This observer would be useless and the Motherland would regard him as a consumer from whom nothing is expected. *We do not wish to play this role in society* : we want our astronomical studies to improve our geography, our roads and our commerce " (cit. by L.C. Arboleda, " Ciencia y nacionalismo en la Nueva Granada en los albores de la revolución de la independencia ", in *Francisco..., op. cit.*, 139-146, p. 142). Doubtless, this and similar texts from which we may quote hint at the need to strengthen the role of the government in deciding political and scientific priorities. Drawing a distinction between both types of knowledge, one, that of the *Republic of Letters*, more susceptible to logic and the other responding to the needs of the *Civil Republic*, entails a new configuration of what Bourdieu calls *scientific field*. This élite that, in the words of Caldas, thinks of itself as antithetical, as " alterity ", aspires to occupy a central place in the public arena due to its assumption that politics was subordinated to science, as history was subordinated to geography : novelties that contributed noticeably to the substitution of the *rhetoric of emancipation* for the *rhetoric of patriotism*.

4. On Zea and his projects, see L.C. Arboleda, " La ciencia y el ideal de ascenso social de los criollos en el virreinato de Nueva Granada ", in A. Lafuente y J. Sala (eds), *Ciencia colonial en América*, Madrid, 1992, 285-319.

spondents dispatched by Mutis already asked the advice of native peoples. The important issue here is, however, that this argument was employed as a throwing weapon to brand Mutis as an elitist and denounce, though subliminally, his alleged incapacity to interface, beyond the narrow circle of his acolytes, with colonial society.

In short, the previous criticisms against the leader of the expedition, did not only cast doubt on his excessive prestige in the colony, but also, as they advocated a new scientific *ethos*, they tended to depict the master as an imperial agent, selfish and despotic, rather than as a patrician sensible to local traditions and in favour of the development of a territory that was already referred to as motherland. Political events will not allow the development of these ideas... ideas that, on the other hand, and after Mutis's death, will prevail in the end as a result of the negotiation forced on the viceroyal authorities by the new generation of Creoles, after the events of 1795 and the banishment of some of its most renowned advocates.

New Spain : motherland and cultural heritage

The studies on Mexican plants conducted by Mariano Mociño and Luis Montaña from 1801 onwards in the Real de Naturales and San Andrés hospitals are one of the most original national episodes of Creole science in the 1700s. It is our belief that the study of this case enables us to explore a dual process : that of the negotiation between Creole and metropolitan scientists regarding the conditions under which Linneus's botany was received, and that of the local appropriation of such ideas through the emergence of scientific procedures stemming from a tradition that professes to be the heir of a cultural heritage[5].

The wards of these prestigious hospitals became a laboratory where the medical effectiveness of the main plants prescribed by native tradition for centuries was to be tested. This, as we shall see later, involved an exercise of caution as vocal animosity towards these practices, branded as primitive or superstitious, was often acrimonious. Without neglecting rigour, the prevailing trend was a scientific pragmatism that, apart from being endorsed by daily experience, was given legitimacy in the studies carried out by Cullen and Brown in Edinburgh. As Alzate formulated it in 1786, the method employed to test the validity of remedies was " to describe their natural history and pharmaceutical preparation "[6], that is to link the climatic conditions and geographical

5. See J.J. Izquierdo, *Montaña y los orígenes del movimiento social y científico de México*, México, 1995, 200-205 ; and also D.B. Cooper, *Epidemic Disease in Mexico City 1761-1813. An Administrative, Social, and Medical Study*, Austin, 1965.

6. J.A. Alzate, " Carta de Edimburgo, 10 de Mayo de 1786 ", *Gacetas de literatura de México* (1791), (vol. 1., ed. Puebla, 1831).

origin with the mixture and dosage of plants employed in each case. The work programme differed, according to Montaña, from " ...the circle of classification and nomenclature typical of cabinet physicists, overloaded with quotations from different authors and always contingent on a positive 'perhaps' ". It was necessary, then, to " ask again for the advice of the herbalist and the *ranchero* and rely on their information "[7].

During the nearly three years that lasted the observation, three overlapping traditions converged. One of them emerged as soon as the first contacts between Mexicans and settlers took place : the medical tradition described in the writings of Cruz y Badiano, Sahagún, Hernández, Jiménez and López ; the second, already mentioned, entailed the re-creation, in the periphery, of techniques stemming from Europe via Edinburgh ; the last one meant to explore the capacity of Linneus's binary model to make predictions. It was, then, an effort to vindicate a medical heritage " dictated by sheer tradition " (as Montaña fittingly puts it), on the one hand, and, simultaneously, to do the only thing which Alzate considered important : to see with one's own eyes the alleged healing properties[8].

The period of public discussion that had begun on the very same day the Royal Botanical Expedition, led by the physician Martín Sessé, left Spain in 1787 thus reached its zenith. As it is known, the pharmacist and, later, professor of Botany, Vicente Cervantes carried in his mind the new classificatory system of Karl Linneus. He had been commissioned to impose it in the viceroyalty by Gómez Ortega (director of Madrid's Botanic Gardens) and, at the latter's behest, fulfilled his task swiftly. What was originally merely a legitimate scientific option would soon become a deep political and institutional clash of interests after the setting-up of the Botany professorship in the newly-founded Garden of Mexico[9].

The argumental display (built upon both factions' empirical and utilitarian convictions) lives off a type of journalism already too mature so as to divide readers into two apparently irreconcilable currents. Polemicists, boasting of their persuasive skills, create the illusion that newspapers are theatres where a drama is performed... a drama that, though cloaked in scientific language, is actually personal. On one side, the members of the expedition playing the role of staunch supporters of Linneus ; on the opposite, a sect of rough and pedantic quack doctors : Cervantes vs. Alzate, Linneus vs. Hernández, New Spain vs.

7. *Discurso de Montaña de 1802*, cit. by J.J. Izquierdo *op. cit.*, 201.
8. For the reception and recreation of Hernandez in the XVIIIth century, see J. Vilchis, " Globalizing the Natural History ", in S. Varey & R. Chabrán (eds), *The World of Francisco Hernández*, Stanford, 1997, 332-258. For Alzate, J.L. Peset, *Ciencia y libertad*, Madrid, 1987, 23-143.
9. Needless to say that, though the most visible, this was not the only battlefield on which the struggle between scientists from the metropolis and those working in the colony took place.

Mexico, science vs. experience, botany vs. natural history... in sum, two Man-icheans in search of their audience and their rhetorical legitimacy[10].

Summing up, a new rhetoric is being built, one which, in contrast with that employed by the like of Elhuyar, Cervantes or other missionaries from the metropolis, did not conform to the propaganda on the usefulness of modernization : it was also aimed at blending the motherland with her cultural heritage within the same reality.

We have already alluded to the priest Alzate, not because he was the wisest or most conspicuous of all Creole polemicists, but rather because he arrogated to himself the role of spokesman. As a whole, the ideas publicized by Gamarra, León and Gama, Santelices, Villaseñor, Bartolache, Guadalajara, Gamboa and others, bring about a sort of *orphean charm*[11], a persistent and melodious rumour capable of bridging the gap between sages and laymen and of combin-ing old knowledge with new disciplines : ethnography, archaeology, medical geography, cartography and anthropology, in pursuit of a national memory that will soon become nationalist.

Actually, Alzate, when giving his support to the nomenclature of traditional herbalism based upon the healing property of plants, was applying the new know how of science (with its hegemony over all other disciplines) to fields typical of political ideology. If this politicization of scientific action is not taken into account the concessions made by both parties in the course of their negotiation will still astonish us. For instance, in the graduation public exams of 1793, Vicente Cervantes will maintain that a student, apart from distinguish-

10. He who wishes to appreciate a perfect example of incisive, scathing and insightful prose (Baroque as far as its style is concerned but enlightened in its content) can read the works reflect-ing this literary and scientific antagonism. To exaggerate, though, would be inappropriate, as the antagonists coincide in their mockery and diatribes against the futile ergotism of the scholastics. Moreover, they are agreed too on their patriotic strivings and their patrician zeal. Something, how-ever, separates ones from the other : Creoles study the new botanical knowledge in the crucible of the Mexican medical corpus. Thus, Alzate cries out ironically : " What is the use of including this or that plant in such genre, or such species, if it has properties very different from those that, because of their similarities, must be included in a given category ? In Europe, experiments bound to fail are conducted as a result of which parsley and hemlock are viewed as similar as far as their organization is concerned. In New Spain, on the contrary, we eat plants and fruits reputedly poi-sonous, were Linneus's botanical legislation to be true... " (J.A. Alzate, *Gacetas de Literatura*, 1788, t. I, 20-22). Since the beginning, the dispute (sprinkled with references to the new chemical nomenclature) is dressed with an unpleasant anti-metropolitan aftertaste or, as Morelos pointed out, with feelings of the nation. The dissemination of the new values that shaped what Clifford Geertz calls *integrating ideology* was the most important event : that is, a self-governing political entity endowed with concepts fraught with authority and legitimacy (C. Geertz, *The Interpretation of Cultures*, New York, 1973, 190-218). To express it differently, we are referring to the values emanating from a kind of eclecticism that combines the political virtues of the Aztec " civilization " with the juridical potentialities spawned by the " sovereignty of sovereignties " of the Baroque period. It is a return to the primeval antiquity, an attempt to re-create an idiosyncratic identity that combines Catholic organicism and the good omens of astrology, the salutary herbal-ism and Anahuac's fertile agriculture.

11. See J.R. Jacob, " 'By an Orphean charm' : science and the two cultures in seventeenth-cen-tury England ", in P. Mack & M.C. Jacob (eds), *Politics and Culture in Early Modern Europe. Essays in Honor of H.G. Koenigsberger*, Cambridge, 1987, 251-273.

ing plants by judging from their gender, will also " ...describe their virtues and will uphold the opinion, in opposition to Linneus's and other authors' claims, that accurate data on their properties can be obtained through well-supervised chemistry "[12].

The mixture of old local wisdom with new European knowledge (not in spite of, but actually thanks to the imperial dynamics of Borbonic despotism) favoured the emergence of a social movement that advocated more autonomy and the spread of a type of creativity that fancied itself as sovereign. Creole initiatives towards assimilation (without giving up a sense of belonging to tradition) and towards re-creation of scientific knowledge and practices refuse to comply with the pattern of simple automatisms characteristic of diffusionism, as well as with that of forcible antagonisms between Europeans and Americans.

Local dynamics within the framework of an imperial background

In the previous pages we have sketched some of the effects that the Hispanic world's inclusion in an imperial structure had on the process of globalization of science. To avoid an entrenched historiographic habit, we have tried to take into consideration the two extremes that imperial reality sets into motion : the metropolis and the colony. Indeed, the endeavour towards *aggiornamento* made from the end of the seventeenth century onwards in the Peninsula, and stepped up after the accession of the new Borbonic dynasty to the throne, will soon be subject to tensions that will forge an institutional identity and a professional *ethos*. Both can be explained if we consider their crystallization into an essential part of an imperial strategy. At the opposite extreme, we find movements that reproduce scientific practices dictated, on the one hand, by the metropolis' demanding character and, on the other, by the colony's patriotic ambitions.

There was not just one Academy, but many academies. Is this a sign of the failure by ruling élites to develop a consultive body located at the top of the pyramid in charge of taking decisions regarding scientific and technical matters ? Does this also show scientists' incapacity to gain legitimacy in an autonomous and self-accounting space ? The answer is, undoubtedly, yes. But

---

12. V. Cervantes, *Ejercicios Públicos de Botánica*, México, 1793, 8. Alzate, in his attempt to reach a new consensus, acted in the same way with regard to Lavoisier's nomenclature. In 1791, he admits it, without wavering (wielding even the same arguments previously employed by Cervantes himself), and writes : " ...I would not dare describe this system as entirely false ". Did he happen to notice the similarity both of its etymological function and the nomenclature of chemistry (the name assigned to a substance describes its process of synthesis) and that of the nahuatl with regard to the native medical tradition (the name of the medicine describes its healing properties) ? Perhaps, why not ? : in both cases we are dealing with a nominalism that assigns a name to that regarded as intrinsic to the nature of a given substance. See A. Saladino, " La Química divulgada por la prensa ilustrada del Nuevo Mundo ", in P. Aceves, *La Química en Europa y América (siglos XVIII y XIX)*, México, 1994, 177-199.

it is also true that scientific models set during the Baroque period (re-creating Renaissance experiences), and likely to have been easily emulated (as was the case in the field of language, history or fine arts) did adapt badly to the requirements imposed by the structuring of the empire (both in the Peninsula and in America) in the course of a process billed by many historians as colonial reconquest. During that century, science and technics were envisaged as privileged tools of the new state and imperial policy. Pressing needs, on their part, turned scientists into government agents, thus providing a foretaste of novelties common in our century and widely experimented during the revolutions that broke out in the American colonies or in neighbouring France. Thus, the late institutionalization of modern science in Spain may also be labelled as more modern due to its statist, utilitarian and instrumentalist drive ; it was also heavily politicized and, if we may resort to an analogy, it suffered from osteoporosis. That is, it was inserted in a system eroded by lack of calcium but which grew up at the same time, apparently healthy but actually fragile. Among the institutional models prevalent in Europe, and in spite of persistent efforts to emulate the former, none seemed to have worked out in Spain. In fact, we cannot help having the impression that local dynamics in the Peninsula led to the assumption of the leading role by the military and had as a concomitant effect the fact that institutions tended to turn into government agencies.

By the end of the 1700s, the conviction that the empire was not viable spread among the Hispanic élites. As many old marriages, it seemed as though the different kingdoms that formed the monarchy could not live together nor split from each other. Whether the flood of complaints was balanced or not, the truth is that the end of a world that vanished like water between one's fingers seemed imminent. Charges against the cultural incompatibility between both parties were soon added to the reproaches heaped upon the metropolis for its cruel exploitation of the colonies. Either because of a vernacular-like mystification, or because of the maturity reached by institutions in colonial centres, the truth is that in Mexico and in Bogotá, our case-studies, the modernizing capacity of the communication channels established between Europe and America by the empire begins to be questioned. It is undeniable that the metropolis was responsible for the introduction of Newton, Linneus and Lavoisier into the colonies : the success achieved not in spite of, but thanks to the aforementioned imperial link. None the less, American intellectuals who criticized the obsolete nature of most of the institutionalized ideas did emerge ; they were partly right since, if such ideas were modern in contrast with the scholasticism pervasive a few decades earlier, they were approaching their expiry date in contemporary Europe. It would be unfair not to acknowledge this point and recognize the rightness of part of Alzate's support of " natural " classificatory systems and Caldas's vindication of botanical geography. To put it briefly, to reduce the nature of the polemics so far outlined to a mere confrontation between Creoles and metropolitans is, after all, a contagious simplification.

The scientific credibility of the opponents was not the only aspect that raised doubts. Criticisms were also fraught with alternative values to those already prevalent. It suffices to recall how they presented the public as opposed to the private, the theoretical as opposed to the pragmatic, the paradigmatic as opposed to the local, academic interest as opposed to patriotic interest or the cabinet study as opposed to field work. It was not an alternative to modernity, but an alternative way to understand modernity. If we were to single out a value that encapsulated the new *ethos*, it would be pragmatism. Creoles did not hesitate to erect the wall of social legitimacy to counter the veracity of the ideas endorsed by European scientific academies. Thus, leading members of the colonial élite try to transfer the scenarios of epistemological validation to America ; they do not give up the experimental verification of principles, rather they adopt new principles claiming that this action would result in better social welfare. They introduce, likewise, epistemological criteria of a lesser rank, incompatible, therefore, with those fostered by the metropolis. They do not promote a non-scientific ideology, rather they wish it to be self-accountable and politically committed. To conclude, it can be said, as we have tried to show, then, that the imperial structure within which colonial science developed did not leave any choice to the élites of New Spain and New Granada but to devise an alternative and distinctive *ethos*.

THE BACKWARDNESS OF PHYSICS IN 19th CENTURY MEXICO

María de la Paz RAMOS LARA - Juan José SALDAÑA

THE TEACHING OF PHYSICS IN THE COLLEGE OF MINING
AND THE SCHOOL OF ENGINEERS

Scientific activity burgeoned considerably in late 19th century New Spain, both in quantity and quality, reflected in broad dissemination of science by journals, periodicals, pamphlets and books in which scientific news and contributions were made known by both Western and New Spanish scientists. These were accompanied by the introduction of instruction in modern science in traditional courses (such as natural philosophy in the context of philosophy courses), and the founding of educational institutions frankly aimed at scientific education.

Both domestic and external features influenced the social and cultural dynamics that favored this increase in scientific interest. External effects included policies implemented by the Spanish Crown and Spanish scientists as well as the influence of European scientists doing research and teaching in Mexico. On the domestic scene, renewed interest resulted from the culmination of a scientific tradition that rose in the 17th century, during which a small scientific community developed out of discussion groups that encouraged and shaped generations who in turn, incorporated, assimilated and cultivated modern science. New Spanish interest in science was both theoretical and practical, though in the late 19th century, practice was predominant, such that the outstanding role of science in New Spanish society occurred around four axes : mining, culture and education, public works, and knowledge of Mexico's territory, its natural wealth and its inhabitants[1]. Of these four, mining, and culture and education were involved in the institutionalization of science.

---

1. J.J. Saldaña, " Ciencia y Felicidad Pública en la Ilustración Americana ", *Historia Social de las Ciencias en América Latina,* México, Ed. Porrúa, 1996, 151-207.

Because of its experimental, systematic and practical nature, physics became a favored cultural and educational discipline among the inhabitants of New Spain. Besides buying books, journals and instruments contributing to the study of the discipline, they began to cultivate it with a variety of objectives. Among these were the application of astronomy for geographical purposes, the teaching of natural philosophy in formal courses, explanation of diverse natural phenomena (eclipses, comet trajectories, aurora borealis, the course of the planets in their solar orbits, and earthquakes), the study of unresolved border problems seeking to propose solutions, study and analysis of scientific explanations and theories, functioning of scientific instruments (the barometer and thermometer for example)[2].

Some in America tried unsuccessfully to institutionalize modern physics. Noteworthy in this attempt were the Jesuits, who even wrote up courses in philosophy to be used as text books, however their efforts came to an end with their expulsion by the King in 1767. Juan Benito Diaz de Gamarra, a professor of philosophy at the College of San Francisco de Sales, also published a book in 1774 entitled *Elementa recentioris philosophiae,* considered to be the first book on modern physics written by a New Spaniard in New Spain. Although his book became very popular, to the extent of being recommended for use at the Royal and Pontifical University of Mexico, the project did not reach fruition due to attacks from his colleagues and denunciations made by them before the Holy Office of the Inquisition. However, institutionalization did take place in the context of mining, thanks to a conjunction of both New Spanish and Peninsular financial interests[3].

By the end of the 18th century, New Spain was enjoying economic growth particularly in the mining, agricultural and crafts sectors. Mining in particular became the most important sector, reflecting the colony's prosperity[4]. This economic improvement led to miners and merchants joining in influential corporations to resolve common problems and promote specific projects. Promotion of viability, operation and even financing came also from Americans[5].

A number of New Spaniards made legal-scientific-technical proposals to the King for the improvement of mining. The most effective was one proposal by Joaquín Velázquez Cárdenas de León and Lucas de Lassaga in 1774, petitioning for the creation of a Royal Mining Tribunal, a Financing Bank and a College of Metals. The tribunal was created in 1777, but the project to create a

    2. E. Trabulse, *Historia de la Ciencia en Mexico*, (Siglos XVI, XVII y XVIII), Consejo Nacional de Ciencia y Tecnologìa México, (Conacyt)/Fondo de Cultura Económica, 1983.
    3. M.P. Ramos Lara, *Difusión e institucionalización de la física en México en el Siglo XVIII*, México, Sociedad Mexicana de Historia de la Ciencia y de la Tecnologìa, Universidad Autónoma de Puebla, 1994.
    4. D.A. Brading, *Mineros y comerciantes en el México Bórbonico (1763-1810)*, México, Fondo de Cultura Económica, 1975.
    5. J.J. Saldaña, *op. cit.,* 154.

college remained without effect at the death of both proponents in 1786. In view of the foregoing, King Carlos III appointed the famous Spanish mineralogist Fausto de Elhuyar as new director, who then came to Mexico City in 1788 and in 1792 founded the College of Mining[6].

While in New Spain it was in the context of mining that physics was institutionalized, in several European countries it also took place in naval and military academies. The need for this was recognized in Mexico after independence from Spain, in view of which the Mexican Constitution of 1824 established that physics be taught in the naval and artillery colleges as well as in colleges of engineering. From its inception, the College of Mining was dependent on the group or party in power. During the Colonial period it depended on the community of miners and after Independence upon the State. With this change, the college broadened its field of activity from mining to public works and the industrial sector. It was also subjected to a number of structural changes which ultimately transformed it into the National School of Engineers in the second half of the 19th century.

Physics and mathematics were considered the most important disciplines in the training of engineers, whence the need to frequently increase the number of professorships in these areas, and to up-date books on physics and instruments and laboratory apparatus. In the late 19th century, such professions as mechanical, industrial and electrical engineer, containing in their *curricula* the greatest number of courses in physics, entered a stage of decline in number of students and job opportunities leading to reappraisal of the teaching of physics. It was decided to increase the practical load in such courses and diminish the theoretical ones, but this did not solve the problem, perhaps because it did not lie in the teaching of physics but in availability of job opportunities, given the fact that trained professionals had a hard time finding work in private enterprises in which the majority of such jobs were filled by foreigners, since when they imported machinery and tools from abroad they also imported trained personnel along with them. This situation was nourished by the fact that Mexico participated on the international stage with a dependent capitalist structure, largely financed by foreign capital. This prevented the establishment of alliances between science, technology and industry in Mexico, as had taken place in Europe and the United States, and which promoted the development of both theoretical and applied scientific activity. As a result, unlike those countries, Mexico experienced no remarkable advances in either theoretical or applied physics[7].

6. Fausto de Elhuyar enjoyed great fame for his discovery along with his brother Juan José of the element wolfram (tungsten).

7. M.P. Ramos Lara, *Historia de la física en México en el Siglo XIX. Las cosas de El Colegio de Minería y la Escuela Nacional de Ingenieros*, doctoral disertation (adviser J.J. Saldaña), Mexico, Faculty of Philosophy and Letters, National Autonomous University of Mexico, 1996. N. Beltraming, *Alianza entre ciencia, tecnología e industria*, México, Anuies, 1977.

EDUCATIONAL AND SOCIAL CONTRADICTIONS GENERATED BY MEXICO'S
DEPENDENT CAPITALISM

Mexico faced a conflict between scientific education and professional activity on the part of engineers, which seemed not to occur in industrialized countries. Recent studies of the social role of the engineer make it possible to visualize the problems he faced in the absence of industries in which to apply his knowledge and in general terms as a result of living in a dependent capitalist country.

Whereas in Mexico the struggle was to build a State that could insert itself into the world's capitalist structure, in Europe, capitalism was expanding via the world market. The dynamics of its industrial expansion rested on control of both the supply of primary goods and the network of trade with the world market's peripheral countries.

From 1821 to 1867, Mexico enjoyed neither economic growth nor equilibrium, due to its grave crises and its shortage of resources[8]. Its subsistence, was based on export of raw materials and import of manufactured goods, though mining and agriculture both suffered serious shortages[9].

The Mexican capitalist state that emerged at the end of the War of Reform, faced social chaos, grave administrative disorganization and a heavy burden of debt. Part of its policy consisted in promoting the activities of private entrepreneurs, stimulating agriculture, transport and commerce ; industry however, remained relegated to a secondary position[10].

Among the economic measures applied were the attraction of foreign capital and the introduction of modern industries and technology. Foreign investment was for the most part devoted to the construction of public works and trade, and did not receive the expected response. With less hesitation than former president Benito Juárez, president Porfirio Díaz made foreign investments and exports the principal aspects of development. Development of the domestic transportation system was promoted to expand the market available to domestic producers, thus preparing the nation for the great era of modern industry[11]. Foreign capital played an important role in domestic savings which, according to Rosenzweig[12], could be viewed as a mine, the owner of which lacked the

8. E. Florescano, M.R. Lanzagorta, " Política Económica ", in *La economía en la época de Juárez,* México, Secretaría de Industria y Comercio, 1972, 75.

9. R. Flores Caballero, " Etapas del Desarrollo Industrial ", in *La economìa en la época de Juárez,* México, Secretarìa de Industria y Comercio, 1972, 154.

10. C. San Juan Victoria, S. Velázquez Ramírez, " La formación del Estado y las políticas económicas (1821-1880) ", in *Mexico en el Siglo XIX (1821-1910),* Ciro Cardoso, Coordinador, Ed. Nueva Imágen, 1981, 65-96.

11. R. Vernon, " Juárez y Díaz ", in *Historia económica de Máxico,* E. Cárdenas, compiler, Mexico, Fondo de Cultura Económica, 1992, 41.

12. F. Rozenzweig, " El desarrollo económico de México de 1877 a 1911 ", in E. Cárdenas, *op. cit.,* 54.

capital necessary to work it, thus foreign entrepreneurs put up their capital and the technical know-how lacking in Mexico and required to develop the country's resources[13]. The relations established between foreign entrepreneurs and domestic elite[14] affected the country's economic, political and social development[15]. Enormous foreign investments linked to substantial exports of raw materials became possible to the extent that Mexican workers were exploited and the ruling class bowed to foreign interests[16] which saw Mexico as a source of raw materials and a market for their manufactures. Hence, most of the investment capital found its way into railroads and extractive industries, principally mining and to lesser extent manufacture[17].

Everything foreign began to abound in Mexico, from machinery to technicians and professionals. The railroads, for example, were built and operated on rails, locomotives and rolling stock, spare parts, iron bridges and supervisory and engineering personnel imported from abroad. Even fuel was sometimes imported as were sleepers and unskilled workers[18]. Industrial plants were usually built by European and US engineers.

Foreigners established the hierarchies in the job market so that the highest posts were always occupied by people from their own countries, including jobs requiring professional support. This caused discrimination against Mexican engineers, though not in all professions, only in those in which foreign competition was unavoidable. Such was the case of mechanical, industrial and electrical engineers. Civil and mining engineers were also affected, though not to the same extent, since they could obtain employment in the government.

Mexico's assumption of a dependent capitalist structure favored serious contradictions of all kinds and in all areas in which science (particularly physics) was involved, that could be industrial, educational and even ideological.

Baldwin emphasizes that it was approximately from 1850 forward that education became a decisive factor in industrial growth and technological change in the West[19]. On the other hand, Mexico's dependent economic structure in the 19th century, deprived the country of the technical change that would pro-

13. M. Wasserman, " La inversión extranjera en México, 1876-1910 : a case study of regional elites ", in E. Cárdenas, op. cit, 268.

14. An elite which, according to López Cámara was largely composed of clerics and large landowners, due to the fact that commercial and industrial bourgeoisie were limited by virtue of being dependent of foreign capital. F. López Cámara, La estructura económica y social de México en la época de la Reforma, México, Siglo XXI, 1967, 210.

15. J. Wasserman, op. cit., 267.

16. R. Vernon, op. cit., 42.

17. F. Rosenzweig, op. cit., 71-72.

18. J.H. Coatsworth, " Los ferrocarriles, indispensables en una economía atrasada : El caso de México ", in E. Cárdenas, op. cit., 222.

19. G.B. Baldwin, " Reflexiones sobre la educación, la tecnología industrial y el progreso económico ", in La Educación en el Mundo de la Industria, Buenos Aires, Angel Estrada y Cía. S.A., 1971, 65-86.

mote industrial and economic growth and at the same time favorably affect education which could not participate as an essential factor precisely because of the absence of an appropriate economic structure. It was therefore impossible for professions devoted to the industrial sector to be successful. By the same token, development of the basic sciences to support the industrial sector, physics among them, did not take place.

However, this did not happen in the case of disciplines such as botany, medicine and geology, theoretical and practical development of which led them to participate on the international scene and even to create institutions subsidized by the government for their development as were : the National Geological Institute (1891) and the National Medical Institute (...). This leads to the conclusion that the fact that Mexico became an exporter of raw materials created conditions for the development of these scientific disciplines, i.e., those linked to the country's natural resources[20]. Furthermore, it impeded full development of disciplines linked to the production of industrial manufactures as in the case of physics[21].

20. Azuela has stated that participation by the scientific community abroad, at conferences, expositions, fairs, publications, *etc.*, favored their incorporation into the dynamics of international scientific activity. Some Mexicans scientists were distinguished by international prizes and recognitions, specially in the fields of biology, geology and medicine. Scientific practice did not cover the entire spectrum of disciplines carried on Europe due to the fact that the link existing with the government limited and conditioned development of research in those aspects relating to expansion of the country's material progress. L.F. Azuela, *La investigación científica en el Porfiriato desde la perspectiva de las principales sociedades científicas,* thesis (adviser J.J. Saldaña), Faculty of Philosophy and Letters, National Autonomous University of Mexico, 1993, 195-208.

21. Trabulse writes that the spectacular advances achieved in medicine, botany and geology were due to the fact that they were fields studied with nationalistic criteria aimed at the study of the country's specific realities. See E. Trabulse, *op. cit.*, Siglo XV, 1983, 173.

LA ENSEÑANZA DE LAS IDEAS NEWTONIANAS SOBRE LA LUZ EN LA UNIVERSIDAD DE CÓRDOBA EN 1782

Carlos D. GALLES

En la ciudad de Washington, cuando corría el año 1912, el eminente historiador Guillermo Furlong se expresaba en una conferencia de la siguiente manera : " No hubo, ni pudo haber, cultura científica durante los largos siglos de estancamiento llamado Epoca Colonial. La modorra imperante y la densa nube de perjuicios que pesaba sobre las diversas regiones del Nuevo Mundo, desde Méjico hasta Buenos Aires y Santiago de Chile, agostaba en ciernes todo conato cultural. Por otra parte, cosa sabida es que la ignorancia era para España una política ya que, gracias a la misma, conservaba su predominio sobre las colonias de América "[1].

Sólo la larga recorrida por los archivos y la compulsa de documentos consiguientes pudieron convencer a Furlong que el suyo había sido un juicio apresurado. Aun en una región del Nuevo Mundo tan olvidada para la metrópolis como la conformada por las tierras que en 1776 pasaron a conformar el Virreinato del Río de la Plata había logrado penetrar la nueva cultura científica.

Es así como Furlong estudia la vida y obra de quien considera el primer matemático argentino de obra importante, el jesuita santafesino Buenaventura Suárez, quien nunca viajó a Europa y cuyos estudios fueron hechos entre Santa Fe, los elementales, y Córdoba, los universitarios. Suárez construye por si mismo sus primeros telescopios, cuadrantes y relojes de péndulo, ayudado por los indígenas de la Reducción de San Cosme y San Damián. Más adelante logra traer de Inglaterra un telescopio de 16 pies y otro de 8 pies, equipados con lentes de Dillon, junto con dos relojes de precisión. Con los resultados de sus observaciones y cálculos publica en 1744 su célebre " Lunario de un siglo ", obra reeditada en tres ocasiones. Nos hemos referido a este primer

1. G. Furlong, *Matemáticos argentinos durante la dominación hispánica*, Buenos Aires, Huarpes, 1945.

astrónomo argentino para ubicarnos en tiempo y espacio y para recordar la gran obra de Furlong quien lanzó una nueva mirada sobre la historia colonial[2].

Debemos señalar que Juan María Gutiérrez, en las antípodas ideológicas de Furlong, había ya destacado la obra de Suárez[3], aunque presentándola como el fruto de un esfuerzo individual, sin apoyo institucional, lo que Furlong refuta con pruebas, en especial la que está constituida por la creación del mencionado observatorio, la cual resultaría imposible sin mediar el apoyo recibido de la Compañía de Jesús.

En la obra de Nota 3 Gutiérrez no logra encontrar ningún ejemplo, salvo la batalla calificada como meramente individual de Suárez, de enseñanza moderna en la Argentina colonial en el campo de la Física. Sus descripciones deben justipreciarse teniendo en cuenta que escribe en plena época de la conformación institucional de la Argentina moderna y para él y los hombres que dirigían el país la cuestión fundamental era acercarse a la cultura científica y a las modalidades sociales de la Europa transpirineica.

Corresponde acotar que esta visión de la cultura científica española no difería demasiado respecto a la que se tenía por entonces en la propia península. Recordemos al respecto lo afirmado por el destacado filólogo Marcelino Menéndez y Pelayo en una obra muy citada : " Verdad es que no apareció en España ningún Galileo, Descartes, Newton, Lagrange, Lavoisier ó Linneo ; confieso de buen grado nuestra inferioridad en esta parte ; no lo da Dios todo a todos ; quizás el terreno no estaba tan bien preparado ; quizá la genialidad española no tira tanto por ese camino como por otros ; quizá la época en que España fue grande y sabia no coincidió con la madurez, sino con los primeros ensayos y tentativas del genio analítico y experimental "[4].

En 1767 terminó la hegemonía de la Compañía de Jesús en tierras americanas al ser expulsados violentamente por orden del rey Carlos III, anticipándose a la decisión del Papa Clemente XIV quien en 1773 decretó lisa y llanamente la supresión de la Compañía. Las consecuencias fueron desastrosas tanto en lo social como en lo económico. En el aspecto religioso y en la educación fueron reemplazados por los franciscanos. Haremos referencia a continuación al curso de Física General dictado en la Universidad de Córdoba por el franciscano Cayetano Rodriguez en 1782, cuyo códice se encuentra en la Biblioteca Central del Colegio del Salvador de Buenos Aires. Se trata de las notas de clase tomadas por Cayetano José Zavala, quien se califica como un " modesto alumno de la Universidad y del Colegio de Montserrat "[5].

2. G. Furlong, *Glorias santafesinas*, Buenos Aires, Surgo, 1929.

3. J.M. Gutiérrez, *Origen y desarrollo de la Enseñanza Publica Superior en Buenos Aires*, Buenos Aires, La Cultura Argentina, 1915.

4. M. Menéndez y Pelayo, *La Ciencia Española*.

5. Se ha utilizado en este trabajo la traducción del latín al castellano de este manuscrito hecha por Celina Lértora Mendoza, así como las notas que lo acompañan. Agradecemos vivamente a la Dra. Lértora el habernos facilitado este trabajo inédito.

Es Rodríguez un singular personaje histórico. Nació en el Rincón de San Pedro, a orillas del Paraná, en 1761 y a los dieciséis años ingresó como novicio en la orden franciscana, siendo ordenado en Córdoba en 1783. Enseñó en la Universidad Teología y Filosofía entre 1781 y 1796. Fue nombrado director de la Biblioteca Pública en 1810, por la Primera Junta de Gobierno, surgida tras la revolución del 25 de Mayo de 1810 con la cual se puso fin a la hegemonía española en el Río de la Plata. En 1816 representó como diputado a Buenos Aires en el Congreso de Tucumán, allí le cupo el honor de ser el redactor de la declaración de la independencia de las Provincias Unidas del Río de la Plata en 1816. Falleció en Buenos Aires en 1823[6].

Si bien Rodríguez fue un gran orador, un destacado periodista, un educador, un político y hasta un poeta considerado en su época, sin dudas no fue un científico, pero su Curso merece un estudio detallado del cual hacemos aquí solo un breve comentario[7]. Nos referimos en particular a la parte que lleva por título " Tercera parte de la Filosofía o Física. Que trata sobre las cosas naturales según las teorías de los modernos ". El libro sexto se refiere a " La Luz, el Lumen y los colores. El cuerpo diáfano y el opaco ". Está dividido en capítulos que responden, según la usanza, a la formulación de ciertas preguntas, éstas son las siguientes :

Cuestión 1. ¿Qué es la luz, qué es el lumen y en qué consisten formalmente?

Cuestión 2. ¿En qué consiste la diafaneidad y opacidad de los cuerpos?

Cuestión 3. ¿Qué debe decirse de la teoría de Newton sobre la luz?

Cuestión 4. ¿En qué consiste formalmente la natura física de los colores considerados en general y en particular?

Cuestión 5. Exposición de los principales aspectos del sistema newtoniano sobre la naturaleza de los colores.

Cuestión 6. ¿Cuál es el origen del color moreno?

Cuestión 7. ¿Cómo se forman los colores del iris?

A lo largo de estas explicaciones no hace Rodríguez mención alguna de los fenómenos a los cuales llamamos hoy en día de interferencia, y que eran ya conocidos en su época, tales como los anillos de Newton o los colores que aparecen en las láminas delgadas[8], pero sí se plantea la posibilidad de la interacción de un haz de luz con otro presentando un *experimentum crucis* en extremo simple y convincente ; el mismo consiste en iluminar una abertura circular en una pantalla opaca con dos fuentes luminosas, observándose que no hay cam-

---

6. Sobre la biografía de Fray Cayetano Rodríguez véase : J.M. Bottaro, *Biografía de un Sampedrino ilustre - Fray Cayetano Rodríguez*, San Pedro, 1896. P. Otero, *Estudio biográfico sobre Fray Cayetano José Rodríguez y recopilación de sus producciones literarias*, Córdoba, La Velocidad, 1899.

7. El códice ha sido estudiado previamente por Celina Lértora Mendoza. Ver : C. Lértora Mendoza, *La enseñanza de la Filosofía en tiempos de la Colonia*, Buenos Aires, Fundación para la Educación, la Ciencia y la Cultura, 1980.

8. M. Blay, *La conceptualisation Newtonienne des phénomènes de la couleur*, Paris, Vrin, 1983.

bios en las iluminaciones producidas del otro lado de la pantalla. La comprobación lleva a Rodríguez a concluir que " la luz no altera a la luz ni impide su propagación ".

La carencia de desarrollos matemáticos en este curso no es por supuesto para nada sorprendente ; los conocimientos en tal campo que Fray Cayetano hubiese podido obtener en sus estudios eran por demás elementales. Al respecto señalemos por ejemplo que Voltaire en su preciosa obra dedicada a las teorías introducidas por Newton utiliza sólo algunos resultados de cálculos y cuando la necesidad de dar una clara explicación matemática de la refracción se hace irremediable, termina apelando a la ironía : *Ce n'est pas ici le lieu d'expliquer en général ce que c'est qu'un sinus. Ceux qui ont étudié la Géometrie le savent assez. Les autres pourraient être un peu embarassés de la définition*[9].

La falta de desarrollos matemáticos en obras de divulgación es común también en nuestros dias, lo cual constituye en verdad la gran limitación de las mismas, el límite insuperable con que tropiezan mientras los conocimientos generales de matemáticas continúen a su actual nivel. En el caso del curso a que hacemos referencia la ley de la refracción de la luz en su aspecto cuantitativo no es ni siquiera mencionada, simplemente se señala que el rayo refractado se aleja o acerca de la normal a la superficie según sea la densidad del medio ; por lo tanto se suministra una descripción de los fenómenos luminosos pero sustentada en el principio de autoridad y no en la clara deducción a partir de primeros principios, de allí la constante necesidad de avalar lo dicho con innumerables citas de autores prestigiosos.

Antes de dar una explicación física del fenómeno del Arco Iris, Fray Cayetano cree necesario sustentar además sus ideas con la autoridad de Virgilio al recordar los versos que comienzan diciendo " Trayendo mil colores diversos contra el Sol ", y también recurriendo a las sagradas escrituras cuando se afirma que Dios dijo a Noé : " Esta es la señal de la alianza que yo contraigo para siempre con ustedes y con todo animal viviente que esté con ustedes para siempre en adelante. Pongo mi arco en las nubes para que sea una señal de mi alianza con toda la tierra "[10].

Esta última referencia no va sin riesgos pues luego de dar la explicación clásica del Arco Iris, el buen fraile se encuentra frente a la necesidad de dar respuesta a la objeción que señala que sus propios argumentos indican que hubo una época (antes del Diluvio, claro está) en que estando el Sol en la posición adecuada, siendo el tiempo lluvioso y habiendo observadores, pues según la Biblia los seres humanos fueron creados con bastante rapidez ; vale decir, dadas las condiciones óptimas para la observación del majestuoso arco en el cielo, este no se presentaba, pues, en una interpretación cruda de las escrituras,

9. Voltaire, *Elémens de la Philosophie de Newton*, Londres, 1738.
10. *La Biblia*, Génesis 9, v 12-13.

sólo comenzó a mostrarse tras las palabras de Dios a Noé. Fray Cayetano responde con mesura que no es así como deben entenderse los versículos, pues el arco iris existió también antes del Diluvio, pero sólo cumplió la significación de una alianza cuando Dios decidió en el momento de su conversación con Noé mostrarlo en todo su esplendor.

La explicación que suministra Rodriguez del majestuoso espectáculo es la bien conocida que está basada en las dos refracciones de la luz en cada gota de agua a la que se suman una o dos reflexiones para dar el arco principal o el secundario. No son citados ni Antonio de Dominis, ni René Descartes, ni Newton, aunque este último va implícito en el discurso pues ha sido mencionado asiduamente en los capítulos precedentes.

Rodríguez menciona asimismo en forma detallada la famosa experiencia del recipiente esférico con paredes de vidrio y conteniendo agua, para ejemplificar su explicación del fenómeno que se da en cada gota. Esta experiencia clave estaba sin dudas al alcance de quien quisiese hacerla, aun en la remota Córdoba del Tucumán, pero no hay constancia de que haya sido hecha, aunque por su sencillez se prestase. Destaca también que es posible formar arcos artificiales, sin esperar la ayuda meteorológica.

La parte más interesante de estas explicaciones está formada por las respuestas que se dan a las posibles objeciones a la explicación teórica. Entre otras aparecen respuestas adecuadas respecto a la posición del sol para que se dé el fenómeno, la interrupción del arco en ciertos casos, el por qué de su forma que se mantiene siempre circular y las características de la formación de colores.

Sin embargo la explicación es claramente incompleta, aún para los patrones de una centuria antes de la época en que Rodriguez imparte su curso. Falta en ella un factor de importancia decisiva tal como es el de dar razón sobre la acumulación de rayos que han entrado por diferentes puntos de la superficie de la gota y que aun así emergen con ángulos similares. Esta circunstancia, que es explicada con sumo cuidado por Descartes y por Newton, no es mencionada. En honor a la verdad, se debe señalar que esta parte de la explicación elemental no es suministrada en muchos de los textos modernos para estudiantes, con lo cual una comprensión del fenómeno del arco iris se hace muy difícil, pues ¿Por qué el ángulo de 42º es determinante en el mismo? Es claro está una pregunta que se formula todo aquel que medite sobre el asunto y para darle respuesta es necesario emplear herramientas de cálculo que eran ajenas a los conocimientos de Rodríguez.

Salvo este importante reparo, pues la falta de matematización del problema físico muestra hasta qué punto la del mismo es meramente nominal, la explicación de Rodríguez sigue el modelo teórico de Newton. Ahora bien, cabe la pregunta ¿Hasta que punto es Rodríguez partidario de las ideas newtonianas en cuanto a los efectos luminosos?

A este respecto cabe indicar que la exposición que hace Rodríguez es bastante exhaustiva pero abunda en críticas no resueltas respecto a la teoría de los colores de Newton. No aborda por otra parte los temas que requieren el uso de conceptos y formalismos matemáticos, se trata de una lectura de Newton, o de textos de lectores de Newton, limitada a ser una descripción cualitativa, lo cual en la óptica puede llevarse a cabo sólo hasta cierto punto.

En su recorrida por varios autores Rodríguez es alternativamente partidario y adversario a lo que llama " los principales aspectos del sistema newtoniano sobre la naturaleza de los colores ". Finalmente concluye que " en la hipótesis newtoniana, según la cual los colores consisten en rayos luminosos intrínsecamente disímiles, y en su diversa refrangibilidad y reflexibilidad, todos los fenómenos del color resultan explicados clara y distintamente ; y esta hipótesis, a su vez, es totalmente acorde con los experimentos ".

Al citar algunas experiencias de Juan Bautista Almeida, en las cuales parece evidenciarse una refutación a la teoría newtoniana, Rodríguez critica la falta de cuidado con que fueron hechas y señala que según " las reglas críticas establecidas para los nuevos experimentos, no deben proponerse como refutación de otros que afirman y fundamentan una teoría, ciertas experiencias no examinadas por los especialistas y académicos del orbe científico, sobre todo si se trata de una sola ".

Conmueve ver en Fray Cayetano tan gran adhesión al método hipotético-deductivo y a la corroboración empírica, con la exigencia de repetibilidad de las experiencias y la necesidad de lograr el apoyo de los pares. No creemos que en su caso fuese el deseo de la pureza epistemológica lo que primase sino más bien era la expresión de su sólido apego al sentido común, tan exhibido luego en su vida pública.

Ciertamente no fue este curso el prólogo a un desarrollo de los estudios en Física en el Virreinato, no lo podía ser de por sí ni estaban dadas las restantes condiciones sociopolíticas para que así fuese. Solo un siglo después la Argentina comenzó a necesitar de profesionales con alguna versacion en las ciencias matemáticas y mucho después pudo recién esbozarse la conciencia en lo imprescindible de la investigación científica para una nación que pretende instalarse en el mundo moderno.

Nollet y la difusión de Newton en el Río de la Plata

Celina A. Lértora Mendoza

Introducción

La comunicación se propone establecer el papel difusor newtoniano que tuvo la obra de Nollet en el Río de la Plata durante la segunda mitad del s. XVIII y comienzos del XIX. La afirmación de su importancia decisiva como difusor del newtonismo se basa en las siguientes consideraciones :

1. La obra de Nollet — sobre todo la traducida al castellano — fue ampliamente conocida en los medios académicos rioplatenses, como lo prueba la existencia de numerosos ejemplares en las bibliotecas, lo que no sucede con otras obras newtonianas y pronewtonianas.

2. Nollet es casi siempre citado como fuente por los profesores cuando explican teorías de Newton.

3. El aspecto de las teorías de Newton que se toma de Nollet (sobre todo la óptica) es el mejor expuesto y el más consensuado.

El marco teórico

El estudio de la difusión de la nueva física en América tiene muchas vertientes. Por una parte se trata de investigar la recepción de las ideas newtonianas, su grado de comprensión, de asimilación y domesticación, además de los eventuales progresos locales en ese orden.

Pero la nueva física se integra también con otros contenidos que en general llamamos postnewtonianos, pero que no son homogéneos en ningún sentido. En líneas muy generales podríamos decir que el postnewtonismo tuvo dos orientaciones temáticas fundamentales : una, la que deriva de su sistema general, otra, la que se centra en problemas particulares, especialmente la Optica[1].

---

1. *Cf.* P.C. C. Abrantes, " Newton e a Física Francesa no século XIX ", *Cadernos de História e Filosofía da Ciência,* serie a, v. 1, 1989, 5-31.

Desde el punto de vista metodológico también podemos señalar dos perspectivas. Una, más abstracta y matematizante, en general es la seguida por la primera vía temática y da por resultado los grandes " sistemas del mundo " de principios del s. XIX. Quizá la figura más representativa de esta dirección en Francia sea Laplace. La segunda perspectiva se acerca más a los principios experimentalistas de la gran corriente homónima del s. XVII (que fue contemporánea de Newton) y desarrolla empíricamente las partes experimentables de la obra newtoniana, en especial la óptica. Esta perspectiva, a la inversa de la anterior que es más homogénea, se presenta variopinta y con diferentes grados de adhesión al sistema total de Newton. Muchos científicos del XVIII entran en esta categoría, y entre ellos nos interesa en especial Nollet.

La razón de esta investigación no ha sido una decisión apriorística sino una exigencia surgida de la investigación sobre la difusión de las ideas de Newton. En el Río de la Plata, esta difusión se cumplió fundamentalmente a través de la cátedra, y por tanto los documentos de confronte son en su mayoria académicos (cursos y conclusiones de exámenes, pues no hubo publicaciones locales sobre el tema). Habida cuenta de que la difusión era sólo informativa, porque no hubo ejercicio científico real en nuestro territorio en la época colonial, la cuestión de las fuentes de información resulta capital.

Estas fuentes pueden ser discriminadas del siguiente modo :

1. Fuentes primarias locales

Son éstas :

a. Fuentes de producción local : los textos conservados de los cursos y de las conclusiones de exámenes.

b. Fuentes de información local : los textos (publicaciones y/o manuscritos) foráneos que estaban en las bibliotecas locales a disposición y consulta de los interesados.

2. Fuentes secundarias : todas las obras de conocimiento indirecto, es decir, mencionadas o reseñadas en las fuentes de información local.

La determinación de las fuentes newtonianas 1.b y 2 y su deslinde, es una tarea un tanto problemática a la que me hallo abocada desde hace tiempo, y he logrado establecer algunos resultados[2]. Por lo que hace a este trabajo, los resultados significativos de este deslinde son :

1. Las obras de Newton no se han hallado en ninguno de los repositodos coloniales conservados ni existen menciones concretas de que hubieran estado en aquella época. Además, las menciones a las teorías newtonianas en los cursos y conclusiones son siempre genéricas (sin cita expresa de lugar) o bien explícitamente indirecte. Por lo tanto hay que concluir que muy probablemente la obra newtoniana fue para nosotros una fuente secundaria, conocida sólo interpósitos expositores.

---

2. V. " Introducción de las teorías newtonianas en el Río de la Plata ", en A. Lafuente *et al.* (ed.), *Mundialización de la ciencia y cultura nacional,* Madrid, Doce calles, 1993, 307-323.

2. Como consecuencia, resulta necesario detectar quiénes fueron los expositores newtonianos de referencia. Para elencarlos tenemos en cuenta las siguientes condiciones : a) que hayan expuesto problemas vinculados a la nueva física, positiva o negativamente, pero en forma concreta y sistemática ; b) que sus obras estén expresamente citadas en las fuentes de producción local ; c) que estas obras tengan algún testigo en las bibliotecas locales. La lista para los requisitos a) y b) es discretamente consistente e incluye los nombres de Jacobo y Juan Bernoulli, Armando Boerhaave, Rogelio Boscovich, Roberto Boyle, Fortunato Brixia, Luis Castel, Claus Eduardo Corcini, Benito Feijóo, José Ferrai de Modoetia, Benjamín Franklin, Guillermo van s' Gravesande, Antonio Genuens, Bertoldo Hauser, Francisco Jacquier, Edmundo Mariotte, Ignacio Monteiro, Pedro van Musschenbroek, Juan Antonio Nollet, José Sigaud Lafond, Vicente Tosca y la colección de las Memorias de Trévoux.

Pero si tomamos en cuenta el requisito de la existencia de testigos locales la lista disminuye bastante. Ya he tenido ocasión de señalar[3] que los cuatro textos académicos más usados en la segunda mitad del XVIII fueron los cursos de Dupasquier, Ferrari, Brixia y Jacquier, de los cuales sólo este último expone adecuada y positivamente la nueva física. Sin embargo, Jacquier parece haber sido de introducción tardía y no fue asimilado sustancialmente en la estructura de los cursos de nuestros profesores, más bien inspirados en Brixia y sus variantes, es decir, en una concepción prenewtoniana de la física empírica. Por eso el sistema newtoniano nunca fue expuesto y aceptado por su propio peso teórico, sino que las adhesiones locales a Newton fueron un tanto erráticas y casi todas se centraron en las cuestiones empíricas.

Nuestros profesores, imbuídos de la tradición experimentalista a la que consideraban digna superación del abstraccionismo del peripato (muy denigrado por casi todos ellos), no pudieron comprender el valor teórico de los conceptos newtonianos y los asimilaron a las " cualidades ocultas " peripatéticas. Eso fue sintomático en el tratamiento del tema de la gravedad. He explicado esto más detenidamente en otro lugar[4].

La mayor adhesión y la mejor comprensión de temas newtonianos se da en los temas susceptibles de comprobación empírica directa y sobre todo en la óptica, y esta situación no es exclusiva del Río de la Plata[5]. Ahora bien, entre las obras de información local de mayor presencia tenemos precisamente la de Nollet. De allí la hipótesis que ha guiado esta investigación.

3. *Cf.* " Bibliografía newtoniana en el Río de la Plata colonial ", en C.A. Lértora Mendoza (comp.), *Newton en América,* Bs. As. ed. FEPAI, 1995, 81-101.

4. " La discusión sobre la gravitación universal en la enseñanza rioplatense (s. XVIII) ", *Sociedad y educación. Ensayos sobre la historia de la educación en América Latina,* Bogotà, Univ. Pedagógica Nac.-Colciencias, 1995, 223-235.

5. Las investigaciones de Patricia Aceves sobre México muestran allá la misma orientación.

LA PRESENCIA DE LA OBRA DE NOLLET EN EL RIO DE LA PLATA

Recordemos que Juan Antonio Nollet (Impré, 1700 - Paris, 1770), tempranamente interesado en las ciencias físicas, estuvo asociado con Dufay al que ayudó en sus investigaciones sobre la electricidad y él mismo trabajó en investigaciones eléctricas y el fenómeno de difusión. Un gran número de sus memorias fueron publicadas en la revista de la Academia de Ciencias entre 1740 y 1767 y sobre sus investigaciones publicó tres obras : *Essai sur l'électricité des corps* (Paris, 1747), *Recherches sur les causes particulières des phénomènes électriques* (Paris, 1749) y *Lettres sur l'électricité* (Paris, 1753). Pero las obras que le dieron celebridad en los medios ilustrados del s. XVIII fueron las de caracter general que constituyeron sus trabajos iniciales : *Programme pour l'idée générale d'un cours de physique expérimentale* (Paris, 1738) y *Leçons de physique expérimentale* (Paris, 1743). Especialmente esta última, traducida al castellano y que gozó de varias ediciones es la que nos interesa. En ella (como en la anterior de 1738) Nollet, más que dar cuenta de sus propias ideas y del resultado de sus trabajos, hace una exposición del estado de la cuestión científico experimental, de modo que sobre todo la segunda, pueden ser considerados verdaderos " manuales de actualización científica ". Esta obra se encuentra en casi todos los repositorios coloniales rioplatenses, en la edición de Madrid de 1757. A ella deben sumarse los trabajos publicados en *Mémoires de Trévoux* desde 1745, pues también existian en la zona.

Esta tesitura también está presente en su discípulo José Aignan Sigaud de Lafond (Bourges, 1730-1810), quien luego de sus estudios médicos se dedicó a la física, siendo sucesor de Nollet en la cátedra de ciencias en el Colegio de Luis el Grande. Como su maestro, se interesó por el fenómeno de la electricidad, al que dedicó algunos trabajos como *Traité de l'électricité* (Paris, 1771) y *L'électricité médicale* (Paris, 1803). Pero en los medios coloniales es más conocido por sus obras generales : *Leçons de physique expérimentale* (Paris, 1767), *Dictionnaire de Physique* (Paris, 1780-1782) y *Dictionnaire des merveilles de la nature* (Paris, 1781). Aunque las citas y menciones son en general de segunda mano, tienen el interés de ser casi contemporáneas y muestran la adhesión local a la interpretación nolletiana de la nueva física.

NOLLET EN LOS CURSOS RIOPLATENSES

Las lecciones de física dictadas por nuestros profesores en la Universidad, el Colegio de San Carlos y los conventos con estudios de grado, durante el s. XVIII, se han conservado muy fragmentariamente, de modo que lo que presento debe ser considerado sólo una muestra. Sin embargo, considero que es válido hacer algunas prudentes extrapolaciones, teniendo en cuenta que los cursos conservados proceden de todos los centros que impartían enseñanza del tema y abarcan un período de más de cuarenta años decisivos para la ilustración americana (1760-1810).

Veremos a continuación los profesores que utilizan a Nollet en sus cursos de física.

Benito Riva (1763)

Este profesor era catalán de origen, habia nacido en 1727 e ingresó a la Compañía de Jesús en 1746[6]. En 1753 está registrado en la Universidad de Córdoba como alumno de Teología y como profesor de Filosofía en el trienio 1762-1764. Murió en Barcelona en 1800.

El códice con sus lecciones de Física, que pertenece a la Biblioteca del Colegio del Salvador de Buenos Aires, parece haber sido copiado por Javier Dicido y Zamudio (el manuscrito carece de portada y el nombre del copista es de otra mano). Es uno de los cursos más extensos y completos que poseemos sobre esta materia, que abarca la física general, la física particular y al final un *Breve tratado sobre el mundo y el cielo*. La Física General se divide en cuatro libros : 1. Principios intrínsecos del cuerpo natural ; 2. Continuo, movimiento y vacío ; 3. Accidente, sustancia y cualidades en general y en particular. La Física Particular Experimental consta de los siguientes libros : 1. Elementos (los cuatro peripatéticos) ; 2. Los mixtos imperfectos o meteoros ; 3. Los mixtos perfectos ; 4. Los fenómenos lumínicos.

En la Parte General, Disputación 1, Secciones 5ª y 6ª trata el sistema de Newton (pp. 19-27) en forma bastante amplia, pues lo explica en 16 proposiciones, rechazándolo por considerar que el concepto de " fuerza de atracción " es inexplicable. La fuente preferida del profesor aquí no ha sido Nollet. Pero también tenemos una razón coherente de este rechazo. En los temas de esta parte, una fuente directa y muy usada por Riva es la obra *Física experimental moderna y sistemática* de Antonio de Herrero, a la que se cita explícitamente muchas veces. De este autor recibe Riva argumentos contra el peripatetismo y su negativa propensión al abstractismo verbal como única explicación física. Lo que pensaba Riva, dicho con palabra de Herrero, está textualmente copiado en la Sección Sexta de la Disputación Tercera a propósito de la generacion de los mixtos, donde se rechaza la existencia de una forma sustancial propia en ellos. A propósito de las afirmaciones peripatéticas se copia en castellano el texto de Herrero en que hace burla de los " terminillos " como " aseidad ", " perrunidad "[7], *etc.* La " gravedad " newtoniana debió parecerle algo semejante, dado su elevado grado de abstracción, cuya diferencia teórica con estos casos seguramente no pudo discernir.

6. Cf. G. Furlong, *Nacimiento y desarrollo de la filosofía en el Río de la Plata (1536-1810)*, Bs. As. Kraft., 1952, 178 y también mi libro *La enseñanza de la filosofía en tiempos de la colonia*, Bs.As. FECIC, 1979, 215 ss.

7. Citado en p. 90, parág. 117-118.

Aunque Riva dedica el último libro de la Física Experimental al tema de la luz, no estudia especialmente a Newton sino que defiende la teoría corpuscular y explica la diferencia de los colores por variación de las vibraciones. Además se dedica bastante extensamente a las leyes de refracción y reflexión, y también a la intensificación de la luz y a la fuente experimental de la luz (el " fósforo " como sustancia inflamable por fricción que provoca resplandor). Tampoco aquí Nollet es fuente especial, aunque e posible que la referencia al fósforo la haya tomado de su Lección XV.

En cambio, resulta interesante observar que Riva se ocupó más que otros profesores de temas experimentales muy actualizados, como la naturaleza y uso de las sales, el magnetismo (que explica siguiendo a Regnault) y la electricidad. En la sección undécima del Libro Tercero (p. 509-512) se explica el fenómeno siguiendo la *De electricitate dissertatione* de Nollet, publicada en las *Mémoires de Trévoux* de 1743. Este mismo curso cita estas Memorias varias veces, pero se dudaba de que hubiesen sido efectivamente consultadas dado que no aparecian ejemplares. En 1988 encontré unos cuarenta volúmnes de esta serie, comprendiendo estos años, en un repositorio no catalogado de la Orden Franciscana. Está confirmada entonces su existencia. Es posible que estos volúmenes hayan pertenecido a la Biblioteca Jesuita de la Universidad que pasó a los Menores luego de la expulsión y que por eso se hayan conservado en sus conventos. La precisión de este parágrafo muestra que la lectura de Nollet resultaba comprensible para nuestro profesor.

Pantealeon Rivarola (1781)

Nació en Buenos Aires en 1757 y se graduó en ambos Derechos en la Universidad de Santiago de Chile[8]. En 1776 ocupó en Chile la cátedra de Leyes. En 1778 se ordenó sacerdote y regresó a Buenos Aires, donde regenteó la cátedra de Filosofía del Colegio de San Carlos en 1779. Como capellán militar participó en la resistencia contra las invasiones inglesas de 1806 y 1807. En 1812 formó parte de la Junta Conservadora de la Libertad de Imprenta. Murió en Buenos Aires en 1821 y fue enterrado en la Iglesia de San Ignacio.

De los cursos dictados por Rivarola sólo nos ha quedado el de Metafísica, en el que se ocupa de una serie de temas físicos en relación con la vida psíquica (parte del tratado *De anima*) : olor, color, sonido, sabor, *etc.* en relación con los sentidos y su fisiología. En la Cuestión Quinta de este tratado se ocupa del sentido de la visión y a propósito de esto enuncia las leyes de refracción y reflexión de la luz inquiriendo al mismo tiempo sobre su naturaleza y a continuación, en la Cuestión Sexta, trata los colores. Es éste el punto en que se refiere a Newton, de un modo totalmente aprobatorio. En especial se ocupa de explicar y fundamentar experimentalmente la teoría de la reflexibilidad y refractabilidad propia de cada rayo heterogéneo, con argumentas y experien-

---

8. G. Furlong, *op. cit.,* 397.

cias de s' Gravesande, Pluche y Nollet. En especial estos dos últimos parecen haber sido consultados directamente para redactar esta parte del curso.

Debemos reconocer que la síntesis que nos da Rivarola del tema de la luz es básicamente correcta y — como dije — contiene parte del material del Tomo 6 de las *Lecciones* de Nollet. Por otra parte, es significativo que Rivarola cita en su curso un elenco muy nutrido de físicos, sobre todo teniendo en cuenta que es un curso de metafísica, en que estos temas no dejan de ser colaterales. Por ej. cita (a veces con mención expresa de lugar) a Malpighi, Tosca, Hauser, Feijóo, Mariotte, Pluche, Corcini, Almeida y otros. Por lo tanto, podemos colegir que Rivarola habría seguido la misma tesitura en el curso de física. Pero si — como sucede en la óptica — se declaraba newtoniano, es algo que no he podido establecer, y tal vez sea poco probable, teniendo en cuanta otros casos.

Cayetano José Rodríguez (1782)

Nació en la ciudad de San Pedro (Provincia de Buenos Aires) en 1761 e igresó a la Orden Franciscana en 1777, ordenándose en 1783. Cuando los franciscanos tomaron la conducción de la universidad (luego de la expulsión de los jesuitas), impulsaron al mismo tiempo la creación de cátedras en Buenos Aires. Rodríguez enseñó en ambos centros. En el colegio de Montserrat (de la Universidad) dictó el Curso de Física General en 1782 y diez años más tarde ya dictaba Teología en el Convento de Buenos Aires. Se sabe que en 1796 escribió las lecciones de lógica, cuyo manuscrito está perdido. Producidos los acontecimientos de 1810 abandonó definitivamente la enseñanza para dedicarse a la política, participando activamente como diputado en la Asamblea del año XIII (que dictó medidas constitucionales para el nuevo estado) y en el Congreso de Tucumán de 1816 que declaró formalmente la independencia[9].

El curso de 1782 (que se conserva en la Biblioteca del Colegio del Salvador) fue copiado por Cayetano José Zavala, quien había ingresado a Montserrat el año anterior y terminó sus estudios en 1788, ordenándose de sacerdote y ejerciendo el ministerio en Chuquisaca[10].

El curso es, como todos, muy ecléctico, y también es uno de los más " modernizados " ya que contiene un considerable acopio informative de temas experimentales, aunque trate de la parte general de la materia. Se divide en 8 libros : 1. los principios generales de los cuerpos ; 2. El lugar y la localización ; 3. El movimiento y la gravedad ; 4. Elasticidad y densidad ; 5. El sonido ; 6. La luz y los colores ; 7. Sabor y olor ; 8. Temperatura y humedad. Apreciamos en esta distribución la idea de tratar los temas más importantes de

---

9. V. mi trabajo *La enseñanza de la filosofía* ..., *op. cit.*, 241 ss. y bibliografía, sobre la obra de G. Furlong p. 245 ss y el *Estudio biográfico sobre fray Cayetano José Rodríguez y recopilación de sus producciones literarias,* Córdoba, 1899, para la actuacion posterior a 1810.

10. *Cf.* G. Furlong, *op. cit.,* 251. nota 2.

la física experimental (que en otros cursos se consideraba " especial ") como " propiedades generales de los cuerpos " y por eso enuncia cada libro de una manera un tanto barroca : " El cuerpo natural en cuando colocado... en cuanto pesado... en cuando sonoro... " *etc.* De este modo puede incorporar una serie de temas que no siempre se trataban en esta parte.

Rodríguez se ocupa de la teoría gravitatoria, rechazándola con el conocido argumento de " la causa oculta ". Pero no la expone donde hubiera sido su lugasr sistemático (en el Libro 3) sino en el 6, como un tema conectado y un presupuesto a la teoría de la luz. La exposición de la luz, en conexión con la teoría gravitacional — aunque parte de un presupuesto correcto — es bastante inconsistente porque en principio acepta la hipótesis lumínica corpuscular y sus efectos, después de rechazar la de la gravedad y haber postulado su conexión. Más adelante, al tratar la teoría de los colores, presenta dos versiones contrapuestas : en la Cuestión Cuarta se rechaza la teoría newtoniana y se acepta la tesis de Tosca, Almeida y otros sobre la diferente textura de la superficie como causa de la diversidad de colores. En cambio, en la Cuestión Quinta ofrece una versión pronewtoniana, que pareciera responder a algún pedido de los alumnos pues indica que desea satisfacer " vuestra curiosidad sobre todos los temas " (p. 179). En este caso menciona explícitamente a Nollet entre los continuadores de las investigaciones de Newton y en efecto la mayoría de los argumentos y la experiencias están tomadas de él, sea directa o indirectamente (seguramente leyó también a Pluche y Almeida, como era hábito común entre nuestros profesores)

El caso de Rodríguez es paradigmático de la perplejidad de nuestros profesores frente al sistema de Newton y hace más clara la hipótesis de que si la parte óptica fue — dentro de lo que cabe esperar — mejor comprendida, no sólo se debió a su caracter más inmediatamente intuitivo, sino a que en este tema se contó con exposiciones más sencillas y asequibles, entre las cuales la de Nollet es la más importante de las exposiciones pronewtonianas.

Elias del Carmen Pereyra (1784)

Este franciscano había nacido en América en 1760, ingresó en la Orden de los Menores en 1776, profesando en Córdoba[11] y fue lector de Filosofía y Cánones en la Universidad de Córdoba de 1778 a 1788 y de Artes y Filosofía de 1783 a 1800. Fue uno de los encargados de la compra de libros para suplir el dictado. Producida la Revolución de 1810, fue Guardián del Convento de Córdoba (1812) donde falleció el 15 de mayo de 1825.

Sobre su enseñanza en Filosofía tenemos tres documentos : las Tesis defendidas en 1786 y 1790 y el Curso de física General de 1784. Las tesis de 1786

---

11. *Cf.* las investigaciones de J.C. Zuretti, " Fray Elías del Carmen Pereyra, profesor de la Universidad de Córdoba ", *Itinerarium*, 2 (1947), 353-271 ; G. Furlong, *op. cit.*, 257-258, y referencias en *La enseñanza de la filosofía...*, 177-178.

fueron sustentadas por Gabino de Sierralta y las de 1790 por Francisco Javier Martínez de Aldunate[12] y versan sobre las cuatro disciplinas del *curriculum* : lógica, metafísica, física general y ética. Las tesis de física, que constituyen un resumen de su Curso, son notablemente eclécticas.

El Curso se estructura en cuatro partes o libros : 1. Los principios generales ; 2. El vacío y el lugar ; 3. Cualidades primarias y secundarias de los cuerpos ; 4. Otras propiedades de los cuerpos. Nollet es citado varias veces, pero no siempre en referencia a Newton y los newtonianos, a los que menciona en general y sin citas específicas. Como sucede con casi todos nuestros profesores franciscanos de la época, su fuente principal es Brixia y es posible que de este manual haya extraído las referencias a Nollet. Pero dado que la bibliografía mencionada por Brixia es muy amplia (y en buena parte inexistente en el Río de la Plata) la elección de Nollet es significativa y sin duda guarda relación con el hecho de que su obra estaba a disposición. Así se ve porque lo recomienda como bibliografía, junto con Brixia, en el tema del vacío y la explicación del fenómeno de los tubos capilares (p. 90, líneas 2-5).

Aunque trata de la gravedad a propósito del movimiento de caída libre, no hay una exposición de la teoría de Newton. En cambio su explicación de los fenómenos lumínicos es discutida en las tres primeras secciones del Libro Cuarto, donde la rechaza. Pero no se discuten los experimentos que aduce Nollet en sus *Lecciones,* lo que hace pensar que no consultó este punto, siendo esto muy posible dado que él mismo reconoce haber estudiado el tema por Almeida.

Mariano Medrano (1794)

Nació en 1767, estudió en el Convento de Santo Domingo, según Furlong, que ha investigado la historia de dicho convento[13]. En 1793 ganó el concurso para suceder a Sebastiani. Tuvo como discípulos a algunas reconocidas figuras posteriores como Manuel Masa, Mariano Moreno, Bonifacio Zapiola, Saturnino Segurola. Participó activamente en los sucesos revolucionarios de 1810 y en la vida política independiente. Filosóficamente Medrano fue un escolástico moderado, con bastantes elementos eclécticos y una decidida acción didáctica a favor de la difusión de la ciencia moderna, al menos tal como él la entendía.

De su enseñanza en el Colegio Carolingio de Buenos Aires conservamos el Curso de Lógica, el de Metafísica y el de Física, éste incluso en dos códices copiados por Bonifacio Zapiola y Saturnino Segurola. Dicho curso se estructura en forma de " Conferencias " que son cinco grandes partes o unidades temáticas : 1. Principios esenciales del cuerpo natural ; 2. El cuerpo natural

---

12. Edición de E. Martinez Paz, " Una tesis de filosofía del s. XVIII en la Universidad de Córdoba ", *Revista de la Universidad Nacional de Córdoba*, 6 (1919), 228-286.

13. G. Furlong, *Nacimiento y desarrollo de la filosofía...*, op. cit, 312, y *La enseñanza de la filosofía....*, op. cit, 117-118.

localizado ; 3. El cuerpo natural móvil, pesado, leve, enrarecido y denso ; 4. El sonido, la luz, los colores y los cuerpos opacos y traslúcidos ; 5. Calor, frío, humedad, sequedad y otros estados de los cuerpos en orden al tacto.

Como vemos, es un programa mixto que incluye temas de física general escolástica, de física general ecléctica y otros (como los de la quinta conferencia) que habitualmente se trataban en la física especial. Cada conferencia está dividida en cuestiones y articulos, y la exposición sigue el modo escolástico aunque concediendo bastante espacio a la explicación en relación a la parte disputativa.

Medrano cita con detalle y precisión un número considerable de fuentes, y es atinado pensar que algunas de ellas, las que existían en los repositorios locales, habrán sido consultadas. Además de los clásicos, algunos teólogos y los tratados escolásticos de Fabri y Ferrari, hay menciones precisas a Torricelli, Barrère, el P. Castel, Duhamel, Hauser, Mariotte, Maignan y Saguens, Lana, Musschenbroek y hasta se menciona una correspondencia P. Castel - Newton sobre el espectro. La mayoría de estas menciones corresponden a las tres últimas partes. Para la elaboración de algunas cuestiones técnicas, Medrano privilegia autores más especializados que los tratadistas escolásticos. Así por ej. cuando trata el vacío en forma empírica, se remite a la obra de Bertoldo Hauser, de la cual hay numerosas citas textuales y a través de ella explica las experiencias de Boyle.

En este marco Nollet es la fuente principal de los temas sobre sonido y color, y es a través de él que se expone la teoría newtoniana de los colores. Sin embargo, las adhesiones del profesor son un tanto erráticas. Así, el tema newtoniano central que aborda, el de la luz y los colores, reconoce otras fuentes además de Nollet. El asunto está tratado en las cuestiones 2ª a 6ª de la Cuarta Conferencia. Es interesante mencionar las conclusiones más importantes que defiende. Con respecto a la naturaleza de la luz y su constitutivo formal, sostiene : 1. La luz primitiva de los cuerpos consiste en el movimiento vibratorio y celerísimo del éter contenido en el cuerpo luminoso (p. 228, parág. 514 del códice Zapiola) ; 2. La luz derivada o lumen es el movimiento vibratorio del éter a partir de la luz primitiva, que se comunica en forma rectilinea y se propaga en forma regular hasta el órgano visivo (p. 229, parág. 516). Sobre la naturaleza de la diafaneidad y la opacidad afirma : 1. La diafaneidad en los cuerpos duros y firmes consiste en la multiplicidad y coordinación de los poros en línea físicamente recta ; en los fluídos en cambio, consiste en que además de la frecuencia de poros, sus partículas ceden facilmente a la luz (p. 256, parág. 565) y 2. La opacidad en los cuerpos duros consiste en que sus poros no están conectados directamente sino intercalados y dispersos ; en los fluídos consiste en que además del desorden de sus poros, sus partículas resisten a la luz (p. 259, parág. 569).

En cambio, al preguntar por la naturaleza física de los colores, tanto en general como en particular, responde : " En el sistema newtoniano, según el cual los colores consisten en radios intrísecamente disímiles, y dotados de diversa refrangibilidad y reflexibilidad, se explican óptimamente todos los fenómenos de los colores " (p. 274, parág. 597). Finalmente, al preguntar cómo se forman los colores del iris responde : " El iris primario se forma por la incidencia de los rayos solares en las gotas de agua de la nube lluviosa que van cayendo y que al incidir en la superficie superior padecen una doble refracción y una reflexión. El iris secundario se forma en las gotas de agua de la nube de lluvia que padecen dos refracciones y otras tantas reflexiones " (p. 329, parág. 681). Luego de pasar revista a la discusión experimental sobre la teoría newtoniana de los colores, Medrano termina la exposición citando (en castellano) la frase de Nollet : " Por más que se diga en contra, hemos de confesar que [la teoría de Newton] es ingeniosa, simple y natural " (p. 299).

Como vemos, las conclusiones están evidenciando por una parte un esfuerzo integrador de fuentes, pero por otra también una desconexión teórica y una falta de comprensión del sentido total de la teoría newtoniana de la luz y el color y por eso se la explica sólo parcialmente pretendiendo armonizarla con otras concepciones de diferente base teórica. Sin embargo, la explicación tomada de Nollet es la más clara, concisa y mejor redactada de toda esta temática. Esto significa que lo que Medrano tomó de Nollet logró captarlo, asimilarlo y trasmitirlo correctamente a sus alumnos. La deficiencia proviene más bien de lo que tomó de otras fuentes y no logró armonizar bien. Por otra parte, esta es una situación habitual, como estamos viendo, en los profesores de la época, y no sólo los rioplatenses.

Fernando Braco, 1796 ó 1797

Este profesor era español de origen, franciscano, llegado muy joven al Río de la Plata. Habría nacido hacia 1767, se ordenó en 1787 y durante 40 años se dedicó a la enseñanza[14]. Desde el punto de vista de sus adhesiones intelectuales, es un ecléctico con profunda simpatía por Descartes. En la estructuración de su trienio filosófico, trata la física a continuación de la metafísica, siguiendo la línea de Wolff, aunque no en el sentido de que sea una " Metafísica Especial ". Estos cursos fueron dictados en el Convento Recoleto de Buenos Aires, que concedía grados y su escribiente fue Raimundo Quintana.

El curso de metafísica se divide en dos partes : Metafísica y Física. La Metafísica se divide en dos subpartes : Ontología y Pneumatología y la Física en General y Especial, aunque de esta última no trata en este curso. La Física General a su vez consta de dos partes : en la primera trata los principios generales de los cuerpos y allí sigue el tradicional orden escolástico de las cuestiones. La segunda parte comprende la mecánica. Aunque en otras partes de esta obra (y de su curso anterior de Lógica) sigue muchas doctrinas aristotélico-

---

14. G. Furlong, *op. cit.*, 231 ss y 506 y *La enseñanza de la filosofía…*, 53-54.

escolásticas, aquí se aprecia una gran adhesión a los modernos e incluso, a pesar de su admiración general por Descartes, en varios puntos concretos prefiere a Gassendi.

En cuanto a las teorías newtonianas se mencionan separadamente en distintas cuestiones. Al tratar la elasticidad de los cuerdos (Cuestión 10ª), expone las ideas de Descartes, Boyle y Newton, suscribiendo la de Hauser. No se menciona a Nollet como fuente y la referencia a Newton es muy general. Más concreta es la Cuestión Undécima, que se dedica a la gravedad. Aquí, luego de una somera explicación, rechaza la teoría de Newton tanto como las de los atomistas a los peripatéticos (p. 268, parág. 723). Resulta par lo menos curioso que le parezcan más " modernas " y adecuadas las propuestas de Lana, Castel a Hauser en cuanto al cometido de " salvar los fenómenos " (p. 271, parág. 732). Aquí la relación con Nollet no es explícita y quizá fue consultado, pero es dudoso, y además se trataría de una lectua muy parcial.

Manuel Gregorio Alvarez, 1798

Fue alumno y profesor del Colegio de San Carlos en el trienio 1796-1799, obteniendo la plaza por concurso al parecer muy lucido[15]. Según Furlong debió inscribirse en el Colegio como alumno a fines de 1773 a principios de 1774. Y en efecto, en el códice que se conserva se lo menciona como alumno del primer Curso de dicho Colegio. Se sabe que también completó los estudios teológicos. El único testimonio de su docencia es el manuscrito que enseguida menciono y carecemos de otros datos sobre su actuación posterior.

El códice que contiene sus lecciones de Física es propiedad particular del Dr. José M. Mariluz Urquijo y coresponde al curso de 1798. El alumno copista fue Jacinto Cárdenas. De este alumno sólo concemos los datos que constan en el *Libro de Matrícula* y por eso sabemos que se inició en los cursos de Gramática en febrero de 1795, en Filosofía en febrero de 1797 y en Teología en febrero de 1800.

El curso de Alvarez es una muestra interesante de los esfuerzos eclécticos por introducir temas de física empírica en un contexto filosófico escolástico. Se divide en tres secciones, de las cuales la primera trata la esencia y principios del cuerpo natural, la segunda sus propiedades y la tercera, que ocupa más de la mitad de la obra engloba bajo el título general de mecánica (el movimiento de los cuerpos) temas tan disímiles como las leyes antiguas y modernas del movimiento, la estática de sólidos y líquidos y el tema de la gravedad. En esta última parte (Cuestión 11ª anteúltima de la sección) se pregunta cuál es la causa de la gravedad, rechazando en brevísimos párrafos tanto la opinión de Aristóteles como las de Gassendi, Newton y los sistemas intermedios, concor-

---

15. En el *Libro de Matrícula del Real Colegio de San Carlos,* Archivo General de la Nación, están las constancias del concurso de 1796.

dando con Teodoro Almeida en que es una propiedad ínsita en los cuerpos desde su creación, por decisión divina[16].

Si bien muchos autores son permanentemente citados, entre ellos Newton, en casi todos los problemas de mecánica y obviamente el de la gravedad, en realidad las fuentes directas consultadas han sido seguramente muy pocas. Explícitamente menciona haber leído en su clase textos de Almeida y de Sigaud Lafond, por donde podemos hallar el nexo con Nollet y las ideas newtonianas. Según Furlong[17] para componer su texto se inspiró — aunque no se dice — en las *Institutiones Philosophicae* de los Lugdunenses, lo que se nota en la redacción de las partes más experimentales de la física. Esta es muy posible, pero con la condición de incorporar también a los dos antes mencionados y también a Nollet. En efecto, Nollet es mencionado explicitamente como fuente en los problemas de la electricidad.

Ahora bien, resulta interesante constatar que Alvarez ha leído a Nollet y a su discípulo Lafond, a los que cita y usa explícitamente, y con los cuales concuerda en los puntos que expone. Sin embargo, el sistema newtoniano, del cual ellos mismos son solidarios, es rechazado categóricamente como la afirmación de un efecto con ignorancia de la causa (p. 141). Alvarez es uno de los tantos casas señalados de incomprensión de la sistemática total newtoniana, y por ello usa y acepta parcialmente desarrollos puntuales apoyados en la teorías general gravitatoria, pero como su fuente inmediata en este punto ha sido otra (Almeida) se niega su pertinencia.

CONSIDERACIONES HISTÓRICO-CRÍTICAS

Los casos que he mencionado, que constituyen a mi juicio una muestra válida de la situación académica rioplatense en cuanto a la difusión de las ideas de Newton y la física postnewtoniana, me permiten consolidar las conclusiones de anteriores trabajos en el sentido de que la parte general de la teoria newtoniana, más abstracta y con mayores requisitos previos de comprensión de la problematica físicomatemática, no pudo ser aceptada porque nuestros profesores sólo tuvieron a su alcance una bibliografía elemental que en su mayoría la rechazaba (salvo Jacquier, que fue poco usado). En cambio, la óptica y otros temas experimentales postnewtonianos tenían una difusion más asequible.

¿Qué conexión podemos establecer entre esta constatación y la utilización de Nollet por parte de nuestros profesores? Mi análisis de la situación se resume en los siguientes puntos.

1. La obra de Nollet tuvo aceptación académica porque, a diferencia de otras, más " literarias ", ya traía una estructura académica que permitía servirse de ella no sólo para aprender el contenido de los temas, sino también para

16. G. Furlong, *Nacimiento y desarrollo de la filosofía...*, op. cit., 343-344.
17. *Op. cit.,* 500.

estructurarlos sistemáticamente. En ese sentido es más útil para un profesor autodidacte (como en esto probablemente lo fueron todos los mencionados) que Pluche, Almeida o Regnault, que son las otras fuentes de información sobre física experimental más usadas.

2. De los seis tomos de sus *Lecciones,* las citas más relevantes se relacionan con la óptica newtoniana, es decir, el tema de los libros 5 y 6 y sobre todo este último. Las menciones a Nollet fuera de este tema o a otras obras suyas son esporádicas. Por lo tanto, la lectura de Nollet parece haberse realizado sobre todo en función de la temática newtoniana.

3. Los profesores que exponen mejor el problema de la luz son sin duda los que directa o indirectamente siguen a Nollet, o a Brixia en lo que tiene de común con él. Esto no significa que todos ellos apoyen la teoria newtoniana — como ya vimos — pero esto puede deberse, en mi criterio, a la tradición de la cátedra. Es decir, sabemos que era costumbre que un profesor siguiera la línea de su maestro o su antecesor, sobre todo si eran amigos. Por esta razón me parece explicable la actitud de Rodríguez que expone primero la tesis en forma negativa (tal vez como propia o como la aprendida de su maestro) y luego la positiva, como exigencia de objetividad para la discusión del tema.

4. Si bien la lectura de Nollet en el tema óptico resultó clarificadora para nuestros profesores, no parece que haya servido ni para suscitar adhesiones a la teoría de la gravitación ni para mejorar su presentación cuando se la expone. En este caso ha sucedido lo mismo que con Jacquier. Mi impresión es que en este punto el prejuicio antiperipatético y la confusión entre el ideal experimentalista y las nuevas direcciones de la física matemática postnewtoniana no permitieron a nuestros profesores un punto de vista más coherente sobre el sistema en su totalidad.

Por todas estas razones considero que el papel de Nollet como introductor de la nueva temática física ha sido importante, pero no para el punto central o eje de la teoría de Newton. Por otra parte, esta incomprensión de la parte más teórica de la física fue luego, decenios más tarde, la principal dificultad con la que chocaron los nuevos estudios universitarios.

LOS *PRINCIPIA* DE NEWTON EN LA NUEVA GRANADA[1]

Luis Carlos ARBOLEDA

El seis de julio de 1801 llegó Alexander Humboldt a Santafé, en donde permaneció durante 63 días antes de continuar su viaje al Ecuador. Este tiempo le bastó para familiarizarse estrechamente con el estado de ebullición cultural que se vivía en la capital del virreinato de la Nueva Granada. Inmediatamente reconoció la impetuosa dinámica que comportaba el movimiento de modernización científica y, por supuesto, los obstáculos estructurales que enfrentaba para poder consolidarse, de parte de fuerzas retardatarias con mucha influencia en la orientación de la vida social. Descubrió que este movimiento se caracterizaba por su espontaneidad y por la fragilidad del incipiente proceso de institucionalización de la enseñanza, consecuencia esta última de una etapa anterior de reformas no completamente saldada. Veamos lo que Humlboldt dejó anotado en su diario de viaje al respecto[2] : " En todas partes oigo hablar de la nueva filosofía, como se denomina aquí la enseñanza de la Moderna física, mecánica y astronomía. La juventud americana se halla impulsada por un movimiento intelectual profundo que ni siquiera se conoce, en España. Aquí todo el iniciado se queja del yugo de la iglesia y del absurdo escolástico y quiere liberarse de las ataduras que los peripatéticos tratan de poner a la razón. Hasta entre los monjes hay reformistas. En vano se prohibió bajo pena de destitución, a los profesores de las escuelas superiores, la enseñanza de esta nueva filosofía, puesto que la juventud siguió estudiándola por su cuenta ".

A continuación narra el incidente que acababa de ocurrir antes de su llegada con relación a la negativa de las autoridades eclesiásticas para que un cierto padre Rojas defendiera el sistema copernicano en el convento de los agustinos.

---

1. Una relación más completa de los resultados de esta investigación histórica se encuentra en mi artículo : L.C. Arboleda, " Sobre una traducción inédita de los *Principia* al castellano hecha por Mutis en la Nueva Granada circa 1770 ", (1993) el cual ha sido publicado por ejemplo en : E. Quevedo (ed.), *Historia social de la ciencia en Colombia*, tomo 2, *Matemáticas, Astronomía y Geología*, Bogotá, Colciencias, 1993, 93-113.

2. H.A. Citado en Schumacher, *Mutis. Un forjador de cultura*, Bogotá, Ecopetrol, 1984.

Al generarse una fuerte controversia, el virrey solicitó a Mutis un concepto. Éste aporta elementos de juicio que recuerdan sus discursos anteriores especialmente la defensa del sistema copernicano en 1773. La fuerza de sus argumentos y la claridad de sus explicaciones sobre la expansión incontenible de este sistema por toda Europa terminan por imponerse, y el padre Rojas puede entonces exponer libremente sus tesis públicas.

La intervención a la que se refiere Humboldt es el documentado informe dirigido por Mutis al virrey Pedro de Mendinueta el 20 de junio de 1801, que efectivamente contribuyó a apuntalar el movimiento de institucionalización de la física newtoniana, originada 40 años atrás cuando el mismo Mutis introdujo su enseñanza en la cátedra de matemáticas del Colegio del Rosario[3]. El testimonio de estos acontecimientos le fue trasmitido a Humboldt en 1801, y ello ha debido causarle una gran impresión, puesto[4] que en el obituario que años después publicará sobre Mutis en la *Biographie Universelle* de Michaud, se refiere en los siguientes términos al sabio gaditano : " Como profesor de matemáticas del Colegio Mayor de Nuestra Señora del Rosario, difundió las primeras nociones del verdadero sistema planetario en Santa Fe. Los dominicos no vieron sin inquietud que " las herejías de Copérnico " profesadas ya por Bouguer, Gogin y La Condamine, en Quito, penetrarán a la Nueva Granada pero el virrey protege a Mutis de los monjes que querían que la tierra permaneciera inmóvil. Estos se acostumbraron poco a poco a lo que llamaban todavía " las hipótesis de la nueva filosofía "[5].

Así pues, Humboldt aparece en estas citaciones como uno de los viajeros europeos más autorizados, que supo valorar desde bien temprano el mérito histórico que le cupo a Mutis en la delicada empresa de casi medio siglo tendiente a aclimatar la racionalidad científica newtoniana en la Nueva Granada. Es en el marco de las tensiones que resultaron de la implantación de la nueva filosofía en las instituciones educativas, que se produjo en este país un hecho histórico muy singular : la traducción de los *Principia* de Newton al castellano.

Debo advertir que se trata de una traducción curiosamente fragmentada, e infortunadamente incompleta. Así, el Libro I fue traducido a partir de la tercera edición latina de 1726, revisada y actualizada por Newton, y el Libro III de

---

3. Publicado con el título " Recapitulación de la doctrina copernicana profesada por el sabio Mutis ", en G. Hernández de Alba, *Pensamiento científico y filosófico de José Celestino Mutis*, Bogotá, Ediciones Fondo cultural cafetero, 1982, 110. Sobre Mutis y la enseñanza de la matemática, también se pueden consultar los siguientes dos trabajos : L.C. Arboleda, *Acerca del problema de la difusión científica en la periferia. El caso de la física newtoniana en la Nueva Granada (1740-1820)*, *Quipu*, vol. 4, 1987, 7-30 ; y L.C. Arboleda, " Mutis entre las matemáticas y la historia natural ", en *Historia social de las ciencias : sabios, médicos y boticarios*, Bogotá, Universidad Nacional de Colombia, 1986, En cuanto a Mutis y la enseñanza de la física, consultar el trabajo de R. Martínez-Chavanz, " La física en Colombia. Su historia y su filosofía ", en : E. Quevedo, *op. cit.*, 1993, tomo 6.

4. Los antecedentes de la localización de este manuscrito inédito se informan en Arboleda (1993), *op. cit.*, 94-95.

5. Michaud, *Biographie Universelle*, t. 29, 1823, 658-662.

alguna versión (probablemente ella misma fragmentaria) de la primera edición latina de 1687. No existe traducción del Libro II y tal vez nunca fue realizada. Pero, en contrapartida, hay una valiosísima traducción de un Comentario al texto del Libro I. El manuscrito está muy bien conservado como documento. Consta de alrededor de 300 folios, tamaño 21 cm x 30 cm, escritos en la caligrafía de Mutis (con la excepción de una pequeña parte) por ambas caras. El todo (unas 160 mil palabras) constituye más o menos la tercera del legajo de papeles matemáticos pertenecientes al Fondo José Celestino Mutis del Real Jardín Botánico de Madrid.

No se ha elaborado una búsqueda exhaustiva en el fondo que permita determinar si algunos folios faltantes en el manuscrito de la traducción se han extraviado o nunca existieron. Aparte de que tal manuscrito durante casi dos siglos ha permanecido desatendido en comparación con los legajos restantes, las sucesivas manipulaciones con fines de publicación de pequeños tratados probablemente han ocasionado que se traspapelen algunos folios. Es posible además que el orden original haya sido alterado como consecuencia del aprovechamiento de las distintas informaciones contenidas en algunos folios del manuscrito. Mutis acostumbraba utilizar en la traducción el anverso y el reverso de otros documentos que le dirigían, como cartas, cuentas, convocatorias a actos académicos, e informes breves. Esta circunstancia tal vez desfavorable para su conservación integral, ha permitido sin embargo ubicar la fecha aproximada de elaboración de la traducción, a falta de cualquier otra información complementaria. Las cartas tienen fechas que se ubican entre junio de 1764 y junio de 1773, con un punto de aglomeración a mediados de 1772. Este dato, sumado a consideraciones que posteriormente expondré sobre el momento histórico en que esta traducción era más viable, permite suponer que fue realizada entre 1772 y 1773.

Veamos una descripción a grandes rasgos de las características de la traducción según su contenido·y procedencia original de fuentes. Como se ha señalado, al trabajar en el Libro I de los *Principia* Mutis utilizó una edición que contenía el texto latino de la edición de Pemberton (la tercera) de 1726, hecha aún en vida de Newton y con su supervisión. El análisis del conjunto titulado "Comentarios de Newton", a todas luces traducido por Mutis del latín con posterioridad al Libro I, facilitó identificar la versión utilizada. Se trata de la famosa edición latina con comentarios de los padres mínimos Leseur y Jacquier.

Es interesante recordar algunos aspectos históricos en cuanto a la significación de esta obra como vehículo de acercamiento de un público culto al contenido matemático de la mecánica newtoniana de los *Principia*. Con anterioridad a la aparición del primer volumen en 1739, la comunidad científica seguía con expectación el proceso de preparación de la edición a cargo de Calandrini. En el volumen de 1737 del *Journal* o *Mémoires de Trévoux*, se publica el prospecto de " la nueva edición de los principios de M. Newton a cargo de dos sabios mínimos del convento de la Santa Trinidad de Roma ". Y en el volumen del año siguiente se anuncia que ya está impreso el primer libro. Igual recepción obtuvo la obra en los más destacados periódicos científicos. Se encuentran reseñas, entre otros, en los *Philosophical Transactions of the Royal Society*, institución a la que los autores dedicaron su obra y de la que ambos serán miembros. En las *Mémoires* y en la *Histoire de l'Académie Royale des Sciences*, de la cual también serán miembros correspondientes, y en el *Journal de Sçavants*. Comparable acogida deben haberle tributado los círculos científicos de Berlín, de Roma y de Bolonia, puesto que hasta donde sé, al menos Jacquier fue hecho miembro de la Academia de Ciencias de San Petersburgo y de las principales sociedades científicas y literarias de Italia.

Los ecos del renombre de los Comentarios obviamente llegaron a España, en donde sin embargo Jacquier se hizo célebre especialmente a partir de los años 1770 por la amplísima divulgación de otra obra suya que fue blanco de todo tipo de polémicas : las Instituciones Filosóficas. En efecto, es bien sabido que los ilustrados y eruditos españoles de los años 1750 no estaban ni mucho menos aislados de estas redes internacionales de información científica. Por el contrario, la presencia de estas obras periódicas en los catálogos disponibles de algunos fondos y bibliotecas de esos años pone presente que estaban al tanto de las novedades y del estado de la literatura científica en los principales centros académicos europeos. Entre todas, las noticias de las *Mémoires de Trévoux* gozaban de una especial autoridad entre los eruditos españoles que con seguridad las tenían como principal fuente de información sobre el progreso de las ciencias y de las artes, y de su enseñanza. Un caso notable es el de Feijoo, como se desprende de la lectura de sus Cartas Eruditas, ante todo las que datan de los años 1750. El benedictino no dudaba en recomendar el uso de las *Mémoires* e incluso explicaba el método de servirse con provecho de las diferentes materias sobre las que versaba tan dilatada obra.

Cualquiera que fuese el vehículo de transmisión, las obras newtonianas como la de los padres Leseur y Jacquier llegaron a España y a sus colonias sin mayor dilación que la razonable dentro de las condiciones prácticas de circulación de toda literatura a mediados del siglo XVIII. Nada permite dudar que fueron manejadas, leídas e incluso aprovechadas en cuestiones particulares o en la formación del gusto por la nueva ciencia entre los miembros de las élites ilustradas. Los pocos pero importantes estudios documentados que se vienen adelantando sobre el problema histórico de la difusión de la ciencia newtoniana en España e Hispanoamérica, así permiten concluirlo. Cosa bien diferente es que a pesar de su temprana difusión, obras como los comentarios a los *Principia* tuvieron que esperar a que se dieran en las periferias receptoras decisivos acontecimientos sociales y culturales, para poder cumplir en esos territorios la función de promoción de una cultura científica moderna, que constituía su razón de ser en los centros emisores. Así debemos entender la aparición de los Comentarios en la biblioteca de los franciscanos en Santafe, aparentemente antes de 1760, aunque su aprovechamiento solo se hizo efectivo en 1770, dentro de un proyecto avanzado de difusión y enseñanza de la física newtoniana en el que Mutis estaba comprometido.

Conviene situar la traducción de Mutis en el contexto de la institucionalización de la física newtoniana en Colombia entre 1740 y 1820. El primero de los tres momentos de este proceso se extiende entre 1740 y 1760. Se presenta entonces un diálogo de la física aristotélica con algunos textos cartesianos y casi~newtonianos que, no obstante, son reinscritos en la cosmología peripatética todavía dominante. El siguiente período está caracterizado por una fuerte penetración de la física moderna, principalmente a través de los textos de los experimentalistas newtonianos 's Gravesande, Musschenbroek, Nollet, Sigaud de la Fond, que Mutis difunde ampliamente en la primera cátedra de matemáticas y física del Colegio del Rosario, entre 1762 y 1766. A partir de entonces se desarrolla un proceso irreversible de enseñanza de la física moderna, ciertamente en medio de conflictos institucionales fuertemente ideologizados y que expresan los intereses contradictorios de grupos antagónicos en la enseñanza. A pesar de todo, poco a poco terminará por imponerse un pensamiento promedio sobre la importancia intrínseca y extrínseca de cultivar la física experimental.

Finalmente, será en el marco de un nuevo proyecto de sociedad en el que se integrará esta enseñanza como parte de un sistema educativo pragmático, de utilidad pública. Tal punto de equilibrio en la institucionalización de la física está representado otra vez por una obra : las Lecciones de Física para los jóvenes del Colegio Mayor Seminario de San Bartolomé, de 1820 y cuyo autor fue José Félix Restrepo, un ilustrado perteneciente a la segunda generación de colombianos que por influjo directo de Mutis se educaron en el gusto por la " nueva filosofía ". Más conocido por haber sido maestro del sabio Francisco José de Caldas, Restrepo ejerció una importante enseñanza de la ciencia moderna en Santafé y en Popayán. Hombre público notable en el período de la construcción del estado republicano (diputado, magistrado, ministro), su magisterio excepcional garantizó en buena medida la línea de continuidad en la difusión de la física moderna entre dos épocas en conflicto. Las Lecciones de Física son el primer texto autóctono de física experimental en Colombia durante la época de la República.

El mismo Restrepo es precisamente uno de los jóvenes que junto a Eloy Valenzuela y, un poco más tarde, Fernando Vergara (catedráticos como el primero de matemáticas y física) se formaron no ya tan sólo en la retórica de los primeros discursos y lecciones mutisianas sobre las ventajas del sistema newtoniano, sino en el estudio más acabado de sus lecciones de los años 1770, en las que el sabio debe haberles trasmitido en alguna medida el meollo matemático-físico de los *Principia*.

Atrás había quedado el joven voluntarioso y soñador de los años 1760, que se empleaba en múltiples actividades y proyectos (aparte de su cargo de médico del virrey), en cada uno de los cuales se veía obligado a demostrar talento y autoridad, inclusive en detrimento de su eficacia y proyección social. Ahora, diez años después, nos encontramos con un Mutis que, al menos en el período de 1770-1776, está más compenetrado con la realidad del país y con la idiosincrasia de sus gentes. Es un Mutis más realista que se compromete sólo con algunos proyectos intelectuales, que aplaza otros con prudencia y tacto y, ante todo, un Mutis que en medio de una sociedad en situación de extremos conflictos sabe crear las condiciones que garanticen la buena marcha de sus opciones personales. También es por supuesto un Mutis de convicciones más arraigadas. En el período de siete años al que me estoy refiriendo, además de profesar con más dedicación que antes la enseñanza de las ciencias matemáticas, de preparar inteligentemente sus proyectos de explotación de minas para los años venideros, y de perseverar en sus exploraciones naturalistas, el gaditano ingresa al clero secular, descubre dos veces la quina en la Nueva Granada, asume vigorosamente la defensa de su prioridad en este descubrimiento y trabaja en la empresa de su explotación y comercialización. Por lo demás lleva hasta un nivel insospechado su polémica anterior sobre el sistema de Copérnico-Newton.

A propósito del mayor compromiso de Mutis con las actividades de enseñanza y difusión científica, hay que tener en cuenta que entre 1769 y 1778, en la Nueva Granada como en España y en otras colonias de América, se adelantaban proyectos de modernización de la enseñanza y contra su control hegemónico por las comunidades religiosas. Más organizada su vida y mejor preparado intelectualmente, Mutis encuentra además un ambiente propicio para proponerse empresas intelectuales de envergadura como esta traducción al castellano de los *Principia*.

Recuérdese así mismo que este movimiento modernizador de la enseñanza se expresa con el compromiso de traducir al castellano obras extranjeras, y de elaborar textos y manuales en lengua materna. En el caso de las ciencias físicas y matemáticas, antes de que Rubín de Celis tradujera la Historia del progreso del entendimiento humano en ciencias exactas de Savérien, en 1775, el jesuita Zacagnini había hecho lo propio con una obra mucho más decisiva, como las Lecciones de física experimental de Nollet, en 1755. Antes de que Bails escribiera los 10 volúmenes de sus influyentes Elementos de Matemáticas, para los guardias de corps de Madrid, en 1756, Lucuce dirigía entre 1756 y 1760 una ambiciosa obra colectiva en matemáticas en la Sociedad Matemática Militar creada en Madrid a instancias del conde de Aranda. Por la misma época el jesuita Wendlingen escribía textos modernos en matemáticas fundamentales, y el también jesuita Cerdá publicaba entre otras obras sus famosas Lecciones de matemáticas, álgebra y aritmética (1758-1760) ".

Los criterios que respaldan este movimiento de traducciones y escritura de textos en castellano no podían ser extraños a un espíritu ilustrado como el de Mutis, muy sensible además a los problemas de la enseñanza de las ciencias en realidades tan peculiares como la Nueva Granada. Recordemos algunas de estas normativas. No se tenía que someter a los pocos individuos con talento para las ciencias, a las dificultades adicionales que suponía su aprendizaje en latín, particularmente en aquellas que exigían mayor aplicación como la física y las matemáticas newtonianas. El tiempo empleado en tener conocimiento solvente en latín como para disponer de un buen entendimiento de las obras científicas escritas en esa lengua, podría emplearse en adquirir conocimientos científicos más útiles. Así mismo, la divulgación de las ciencias en lengua materna contribuía a romper el monopolio autoritario del saber detentado por aquellos que disputaban en esa lengua, aunque no dispusieran necesariamente de la vocación para cultivarlo, ni del talento para engrandecer tal saber. Ahora bien, si la difusión y la enseñanza de las ciencias en castellano contribuían directamente al progreso de su entendimiento, al menos favorecían una más rápida toma de conciencia de los aficionados y practicantes sobre su desarrollo en los grandes centros y el atraso de su penetración en nuestros propios países.

Lo dicho anteriormente es suficiente para confirmar una vez más que las periferias participaron desde bien temprano en el proceso internacional de modernización científica, inclusive en los niveles más avanzados. Este es el caso de la primera traducción inédita al castellano de los *Principia* : un testimonio elocuente del empeño con el que se asumió la difusión de la ciencia moderna en la Nueva Granada.

Este documento sui generis, esta traducción fragmentaria e incompleta, es una prueba incontestable de los esfuerzos tenaces que se hicieron por explotar como fuera los recursos disponibles y aprovechar las limitadas condiciones institucionales y sociales de la Nueva Granada en la segunda mitad del siglo XVIII, con el fin de aclimatar la obra paradigmática de la nueva racionalidad científica. Al mismo tiempo, el manuscrito nos invita a rescatar la personalidad de Mutis de una cierta historia del pasado periclitado en la que no se hace más que repetir elogios y lugares comunes, y restituirle el papel activo que desempeñó en la formación de la cultura científica colombiana.

PART THREE

EAST ASIA

CHRISTOPHER CLAVIUS AND LI ZHIZAO

FUNG Kam-Wing

CHRISTOPHER CLAVIUS'S MATHEMATICAL CAREER

Christopher Clavius (Christophorus Clavius, 1538-1612), one of the most celebrated mathematicians and astronomers of his age, spent his early life in his hometown Bamberg, a place of distinctive mathematical community as well as a minor centre for publication of astronomical almanacs and manuals for the usage of astronomical instruments like astrolabes[1]. He joined the Society of Jesus in Rome in 1555 and later studied for several years (1556-1560) at the Jesuit College in Coimbra (Portugal), where he observed a total eclipse of the sun on 21 August 1560[2]. Clavius attended the philosophical lectures of the Portuguese Jesuit Pedro Fonseca (Pedro Alfonsequa) and at the same time it is likely that he was influenced by the famous Portuguese mathematician and cosmographer Pedro Nuñez (1502-1578) who was teaching mathematics in Coimbra (1537-1578)[3]. Clavius began to teach mathematics for the philosophy programme at the *Collegio Romano* in Rome in 1563 when he was still a scholastic in his second year of Theology, and for the rest of his 48 years' life, excepting for several years, he was a member of the Faculty as Professor of

1. On Clavius's biography, the definitive work is J.M. Lattis's *Between Copernicus and Galileo : Christoph Clavius and the Collapse of Ptolemaic Cosmology,* Chicago, London, University of Chicago Press, 1994, 1-29. Also refer to E. Zinner, *Deutsche und Niederländische astronomische Instrumente des 11-18 Juhrhunderts,* Munich, C.H. Becksche Verlagsbuchhandlung, 1956, 154-155 ; K. Takeda, " Sugaku-shi jo no kurabiusu no chi-i (The Position of Clavius in the History of Mathematics) ", in *Kagakushi kenkyû (Journal of History of Science,* Japan), 41 (1957), 1-4 ; E. Knobloch, " Sur la vie et l'oeuvre de Christophore Clavius ", *Revue d'histoire des sciences,* 41 (1988), 331-356.

2. H.L.L. Busard, " Clavius, Christoph ", in C.C. Gillispie (ed), *Dictionary of Scientific Biography,* New York, Charles Scribner's Sons, 1970-1980, vol. 3, 311-312 ; C. Clavius, *Christophori Clavii Bambergensis In Sphaeram Ioannis de Sacro Bosco commentarius,* Romae, Apud Victorium Hehanum, 1570, 295 ; also refer to U. Baldini, " Christoph Clavius and the Scientific Scene in Rome ", in G.V. Coyne, S.J., M.A. Hoskin and O. Pedersen (eds), *Gregorian Reforrn of the Calendar : Proceedings of the Vatican Conference to Commemorate its 400th Anniversary 1582-1982,* Vatican City, Pontificia Academia Scientiarvm, Specola Vaticana, 1983, 137-169, esp. 144-145 and 160-161.

Mathematics (1563-71, 1575-76, 1577-1584, 1587-1595) or as Scriptor (1596 ?-1612)[4]. The philosophy *curriculum* at the *Collegio Romano* comprised three years' studies, with Introduction and Logic occupying the first year, Physics, Cosmology and Astronomy the second, and Metaphysics, Psychology and Ethics the third[5]. A course of mathematics ran parallel with philosophy. In this particular *curriculum*, mathematics was by no means treated as a branch of insignificant educational value. Clavius then wrote an autograph treatise *Modus quo disciplinae mathematicae in scholis Societatis possent promoveri* (The way in which the mathematical disciplines could be promoted in the schools of the Society) to present his pedagogic ideas on the teaching of mathematics and its relationship to philosophy as well as setting up a mathematical school at the *Collegio Romano*.

Clavius declares that : " First a master must be chosen with uncommon erudition and authority ; for if either of these is absent the pupils, as experience shows, seem unable to be attracted to the mathematical disciplines... It also seems necessary that the teacher should have a certain inclination and propensity for lecturing on these sciences, and should not be taken up with many other occupations ; otherwise he will scarcely be able to help his pupils. Now

3. J.M. Lattis, *Between Copernicus and Galileo*, 16. On Pedro Nuñez, refer to J.M. López de Azcona, " Nuñez Salaciense, Pedro ", *Dictionary of Scientific Biography*, vol. 10, 160-162. Pedro Nuñez was the author of several astronomical and mathematical works including *Página do Tratado da Sphera*, Lisboa, 1537 ; a commentary on the *Sphere* of Johannes de Sacrobosco), *De Crepusculis liber unus*, Lisboa, 1542 ; a detailed description about Nuñez's method of circular scale subdivision by using an " nonius ", an instrument consisting of 44 concentric auxiliary circles, was attached to an astrolabe for measuring fractions of a degree. It is noteworthy that Clavius later simplified Nuñez's method in 1593, to reduce the auxiliary circles to one divided into sixty-one parts and divided the limb of the astrolabe into sixty, refer to *Christophori Clavii Bambergensis Opera mathematica*, Moguntiae, Sumptibus Antonii Hierat, 1611-1612, t. II *Geometria Practica*, 10-12, *De Arte Atque Ratione Navigandi*, Coimbra, 1546 (researches on spherical geometry), *Opera quae complectuntur, primum duos libros*, Basel, 1566 (theory of loxodromic curves [Pé-táng, n° 2342]) and *Libro de Algebra en Arithmetica y Geometria*, Antwerp, 1567.

4. *Between Copernicus and Galileo*, 15, 24 ; E.C. Philhps, S.I., " The Correspondence of Father Christopher Clavius S.I. preserved in the Archives of the Pont. Gregorian University ", in *Archivum Historicum Societatis Iesu*, 8 (1939), 193-222 ; W.A. Wallace, *Galileo and His Sources : The Heritage of the Collegio Romano in Galileo's Science*, Princeton, Princeton University Press, 1984, 7 ; F.A. Homan, S.J., " Christopher Clavius and the Renaissance of Euclidean Geometry ", in *Archivum historicum Societatis Jesu*, 52 (1983), 78-96 ; E. Knobloch, " Sur la vie et l'oeuvre de Christophore Clavius (1538-1612) ", in *Revue d'histoire des sciences*, 41 (1988), 331-356 ; L. Maierù, " ... in Christophorum Clavium de Contactu Linearum Apologia' Considerazioni altorno alla polemica fra Peletier e Clavio circa l'angolo de contatto (1579-1589) ", in *Archive for History of Exact Sciences*, 41 (1991), 115-137 ; C. Jami, " From Clavius to Pardies : The Evolution of the Jesuit's Teaching of Geometry in China (1607-1723) ", (unpublished draft, July 1996).

5. R. Schwickerath, S.J., *Jesuit Education : Its History and Principles, viewed in the light of Modern educational problems*, Vienna, B. Herder/Freiburg im Breisgau, 1903, Chapter 4 " The Ratio Studiorum of 1599 ", 107-143, esp. 131-132 ; A.C. Crombie, " Mathematics and Platonism in the Sixteenth-Century Italian Universities and in Jesuit Educational Policy ", in Y. Maeyama and W.G. Saltzer (eds), *Prismata : Naturwissenschaftsgeschichtliche Studien*, Wiesbaden, Franz Steiner Verlag, 1977, 63-94, esp. 66-71 ; A. Scaglione, *The Liberal Arts and the Jesuit College System*, Amsterdam, John Benjamins Publishing Company, 1986, 87-91.

in order that the Society should have capable professors of those sciences, some men should be selected apt and capable for carrying out this task who may be instructed in a private school (*academia*) in various mathematical subjects ; otherwise it does not seem possible that these studies should last long in the Society, let alone be promoted ; although they are a great ornament to the Society and are very frequently the subject of discussion in colloquia and meetings of leading men, where they might understand that our members are not ignorant of mathematical matters... I do not mention the fact that natural philosophy without the mathematical disciplines is lame and incomplete, as we shall show a little later. Secondly then, it is necessary that the pupils should understand that these sciences are useful and necessary for rightly understanding the rest of philosophy, and that they are at the same time a great ornament to all other arts, so that one may acquire perfect erudition ; indeed these sciences and natural philosophy have so close an affinity with one another that unless they give each other mutual aid they can in no way preserve their own worth. For this to happen, it will be necessary first that students of physics should at the same time study mathematical disciplines ; a habit which has always been retained in the Society's schools hitherto. For if these sciences were taught at another time, students of philosophy would think, and understandably, that they were in no way necessary to physics, and so very few would want to understand them ; though it is agreed among experts that physics cannot rightly be grasped without them, especially as regards that part which concerns the number and motion of the celestial circles (orbes), the multitude of intelligenes, the effects of the stars which depend on the various conjunctions, oppositions and other distances between them, the division of continuous quantity into infinity, the ebb and flow of the sea, winds, comets, the rainbow, the halo and other meteorological things, the proportions of motions, qualities, actions, passions and reactions *etc.* concerning with calculators write much "[6].

The suggestion of Clavius was adopted in the Jesuit teaching programme when Matteo Ricci (1552-1610) entered the *Collegio Romano* in September 1572. According to the *Monumenta Paedagogica Societatis Jesu,* Logic was taught in the first year philosophy[7] : [In the second year of philosophy] : the

6. Autograph, " Manu P. Christophori Clavii ", in *Monumenta Paedagogica Societatis Jesu* (Madrid, 1901-1902), 471-473. Also refer to A.C. Crombie, " Mathematics and Platonism ", 65-66 ; C. Lewis, *The Merton Tradition and Kinematics in Late Sixteenth and Early Seventeenth Century Italy,* Padova, Editrice Antenore, 1980, 74-75 ; P. Dear, *Discipline & Experience : The Mathematical Way in the Scientific Revolution,* Chicago, London, The Unversity of Chicago Press, especiahy Chapter 2 " Experience and Jesuit Mathematical Science : The Practical Importance of Methodology ", 32-62.

7. *Monumenta Paedagogica Societatis Jesu* (Madrid, 1901-1902), 478. Also refer to H. Bernard, S.J., *Matteo Riccis Scientific Contribution to China* (translated by E. Chalmers Werner), Westport, Hyperion Press, 1973 reprint, 26-31.

first four Books of Euclid during four months approximately[8], Practical Mathematics one month and a half[9], *Sphere* two months and a half[10], Geography two months[11], and, during the remainder of the year, Book 5 and 6 of Euclid.

[In the third year of philosophy] : Astrolabe two months[12], Theory of the Planets four months[13], Perspective three months[14], during the remainder of the time Clocks and ecclesiastical Computation[15].

Ricci was then learning mathematics, geography and cartography from Clavius during the period of 1572-78. One may safely assume that Ricci was one of the faithful listeners among Clavius' students in the *Collegio Romano.* As Ricci began to preach Christianity in China in 1583, he was able to use to great advantage what he had learned from his master Clavius. It is worthy noting that the influence of Clavius was evident in the preliminary Jesuit *Ratio Studiorum* of 1568 in Rome and in the final version of 1599 in Naples[16].

In 1570, Clavius published the first edition of the *In Sphaeram Ioannis de Sacro Bosco Commentarius* (Commentary on the *Sphere* of Sacrobosco, [Pé-táng, 1585 edition, n° 1308]) in Rome. As the author of a number of substantial texts, Clavius was a dissenter of the Copernican theory and insisted upon

8. Texts being used could include Franciscus Flussates Candalla (François de Foix, Comte de Candale, 1502-1594)'s *De solidorum regularum comparatione (1566)* and Clavius's teaching notes that was believed to be incorporated into his Latin version of *Euclidis elementorum libri XV (Elements of Euclid),* Romae, 1574 [Pé-táng 1591 edition, n° 1297]. See also T.L. Heath, *The Thirteen Books of Euclids Elements,* New York, Dover, 1956 reprint, vol. 1, 104-105.

9. Texts could include Clavius's teaching notes or drafts that was published later as *Epitome Arithmeticae Practicae* (Romae, 1583, [Pé-táng, 1585 edition, n° 1296]) and translated into Italian (*Arithmetica Prattica,* 1586). See also D.E. Smith, *History of mathematics,* New York, Dover, 1958 reprint, vol. 1, 333-335.

10. Texts could include the *Sphere* of Johannes de Sacrobosco, Oronce Fine's (1494-1555) *De Mundi Sphaera* (1542) and Clavius's newly published *In Sphaeram Ioannis de Sacro Bosco Commentarius,* Romae, 1570. On Sacrobosco, see L. Thorndike, *The Sphere of Sacrobosco and Its Commentators,* Chicago, The University of Chicago Press, 1949, 1-75. On the biography of Oronce Fine, see E. Poulle, " Fine, Oronce ", *Dictionary of Scientific Biography,* vol. 15, 153-157.

11. The course in Geography could include Ptolemaic geography or Johann Schöner of Nürnberg's (1477-1547) cosmographical globe, i.e. *De usu globi terrestris,* 1551. See also E.L. Stevenson, *Terrestrial and Celestial Globes : Their History and Construction,* New Haven, Yale University Press, 1921, 82-88.

12. Texts could include Johannes Stöffler's (1452-1531) *Elucidatio Fabricae Ususque Astrolabii* (Tübingen, 1511, [Pé-táng, 1553 edition, n° 2876]) and teaching notes prepared by Clavius which was believed to be the draft of *Christophori Clavii Bambergensis Astrolabium* (Romae, 1593, [Pé-táng, n° 1291]).

13. Texts could include Georg Peurbach (1423-1461), *Theoricae novae planetarum* (Paris, 1543, [Pé-táng, n° 2431]). See also C. Doris Hellman and N.M. Swerdlow, " Peurbach Georg ", *Dictionary of Scientific Biography,* vol. 15, 473-479.

14. Texts could include Georg Hartmann's (1489-1564) *Perspectiva Communis* (Nuremberg, 1542, [Pé-táng, n° 1872]) or Jacques le Fèvre d'Estaples's (Jacobus Faber Stapulensis, 1455-1536), *Introductio Jacobi fabri Stapulensis, in Arithmeticam Diui Seuerini Boetij (1507).* See also L.B. Ritvo, " Hartmann, Georg ", *Dictionary of Scientific Biography,* vol. 6, 143-145.

15. Texts could include Francesco Maurolico's (1494-1575), *Computus ecclesiasticus in summam Collectus* (Venice, 1575, [Pé-táng, n° 2212]). See also A. Masotti, " Maurolico, Francesco ", *Dictionary of Scientific Biography,* vol. 9, 190-194.

16. Lattis, *Between Copernicus and Galileo,* 32.

the geocentric system of Ptolemy[17]. However, for the maximum declination of the sun (*maxima solis declinatio quid, et quanta sit*), Clavius cited Copernicus' measurement in Chapter III of the In *Sphaeram Ioannis de Sacro Bosco Commentarius*[18]. The year 1581 saw the publication of *Treatise on Gnomonics* (*Gnomonices libri octo*, Romae, 1581 [Pé-táng, n° 1301]). Apart from his mathematical computation, Clavius was also closely involved in the Gregorian Reform of the Calendar in 1582[19]. By the time Clavius continued to apply himself in his scientific pursuits, his several influential works were later brought by missionaries or sent by himself to China, and these valuable books were preserved in the Jesuit Nan-t'ang Library at Peking in Late Ming and later moved to Tong-t'ang and Pé-t'ang Libraries in Early Qing. Henri Bernard points out that Matteo Ricci owned a small library in Shaozhou of northern Guangdong between the years 1590 and 1595, during which at least three works of Clavius, his *Epitome Arithmeticae Practicae, Euclidis Elementorum Libri XV* and *Gnomonices Libri Octo*, were kept[20]. Ricci himself also expressed his gratitude to Clavius after receiving two copies of *Astrolabium* in a letter dated 25 December 1597 at Nanchang in Jiangxi Province[21]. Several years later, in 9 May 1605, Ricci finished his translation of Clavius's *Calendarium Gregorianum* (Pé-táng, 1603 edition, n° 1306)[22].

HOW LI ZHIZAO WAS ATTRACTED TO EUROPEAN SCIENCES

Ming scholar-official Li Zhizao (1565-1630, baptized Leon in 1610) first called on Matteo Ricci in 1601 during Ricci's second visit to Peking. From then on, they developed an enduring relationship. Ricci describes Li's personality and his yearning for western learning as follows :" Li Wocun [i.e. Li Zhizao] is from the city of Hangzhou in the province of Zhejiang. At the time I first arrived in Peking, he was a high official in the Ministry of Works and was a doctor of great intelligence. (Highly placed in the *jinshi* examination of 1598, Li was called a " doctor " by the Jesuits) In his youth he produced a *Description of All China* [*Tianxia zongtu* ?] with the fifteen provinces shown in great detail ; he thought it was the whole world. When he saw our *Universal*

17. *Ibid.*, 86-105.

18. " Nicolaus Copernicus eandem pronunciauit grad. 23. minut. 28. secun. 20 ", *In Sphaeram Ioannis de Sacro Bosco Commentarius* (1570), Caput Secundum, 330 ; E. Knobloch, " Christoph-Clavius Ein Astronom zwischen Antike und Kopernikus ", in K. Döring and G. Wöhrle (eds), *Vorträge des ersten Symposions des Bamberger Arbeitskreises " Antike Naturwissenschaft und ihre Rezeption "* (AKAN), Wiesbaden, Otto Harrassowitz, 1990, 113-140.

19. U. Baldini, " Christoph Clavius and the Scientific Scene in Rome ", 146-154.

20. H. Bernard, *Matteo Ricci's Scientific Contribution to China*, 44.

21. M. Ricci, " Al P. Cristoforo Clavio S.I. a Roma (Nanchian, 25 December 1597) ", in P. Pietro Tacchi Venturi S.I. (eds), *Opere Storiche del P. Matteo Ricci*, Macerata, Premiato Stabilimento Tipografico, 1913, Volume Secondo *Le Lettere Dalla Cina*, 241-243.

22. M. Ricci, " Al P. Fabio de Fabj S.I. a Roma (Pechino, 9 Maggio 1605) ", *Opere Storiche*, Volume Secondo, 261-268.

World Map [*Shanhai yudi quantu*], he realized how small China was compared to the whole world. With his great intelligence he easily grasped the truths we taught about the extent and sphericity of the Earth, its poles, the ten [concentric] heavens, the vastness of the sun and stars compared to the Earth, and other things which others found so difficult to believe. From this a close friendship developed between us, and when the duties of his office allowed it, he liked to learn more of this knowledge (*questa scientia*) "[23].

Li was so impressed by Ricci's *Universal World Map* and pressed Ricci to republish the third edition in a considerably larger size. The diagrams representing the nine/ten celestial spheres (so called *Jiuchongtian tu* but named as *Qiankunti tu* in Ricci's Chinese work *Qiankunti yi*, 1605), armillary sphere (so called *tiandi yi* but named as *hunxiang tu* in *Qiankunti yi* and Li Zhizao translated work *Hungai tongxian tushuo*) and other explanatory notes were basically translated from the extracts of Clavius's *Sphaera*. In a letter dated 12 May, 1605 in Peking to Claudio Aquaviva (1548-1623), the general of the Society of Jesus (1581-1615), Ricci wrote about Li learning mathematics from him and conducting researches on Clavius's works as well as making instruments.

" Yesterday, he [i.e. Li Zhizao] sent me from his new official residence in Zhangqiu of Shandong Province, two stone sun-dials, the one plane, the other fixed on a wall, and, which complicated the problem, the wall was not oriented to face the south, but was salant ; thus, to make a still more surprising object, he constructed this inclined, or rather deviating dial ". This mandarin [i.e. Li Zhizao] had studied with me for some years past in Peking on many mathematical subjects. The name of Father Clavius is already known to many Chinese... My student Li Zhizao sent me illustrations and tables under my direction, please send one copy to Father Clavius who would find it interest and familiarity[24].

23. P.M. D'elia S.I. edited, *Fonti Ricciane*, Roma, La Libreria Dello Stato, 1949, vol. 2, 168-172 ; L.J. Gallagher, S.J., *China in the Sixteenth Century : The Journals of Matthew Ricci, 1583-1610*, New York, Random House, 1953, 397 ; F. Hao, *Li Chih-tsao yen-chiu* (Research on Li Chih-tsao), Taipei, Commercial Press, 1966, 19-30 ; W.J. Peterson, " Why did they become Christians ? Yang T'ing-yün, Li Chih-tsao, and Hsü Kuang-ch'i ", in C.E. Ronan S.J. and B.B.C. Oh (eds), *East meets West : the Jesuits in China, 1582-1773*, Chicago, Loyola University Press, 1988, 129-152 ; P. Yuen-sang Leung, " Li Zhizao's search for a Confucian-Christian Synthesis ", in *Ming Studies*, 28 (Fall, 1989), 1-14 ; C. Wanru and H. Shaogeng, " You-guan Li Zhizao zhizuo diqiuyi di shiliao (Historical Materials about the Terrestrial Globe made by Li Zhizao) ", in *Zhongguo keji shiliao* (China Historical Materials of Science and Technology), vol. 12 (1991), 85-87 ; C. Min-sun, " Li Zhizao xinfeng tianzhujiao di yuan-you tankao (An Investigation on Li Zhizao's Belief in Christianity) ", in Hsu Cho-yun *et al.*, *Zhongguo tushu wenshi lunji* (Collected Essays on Chinese Books and Culture), Beijing, Modern Publisher, 1992, 411-420.
24. M. Ricci, " All' Alvarez a Roma (Pechino, 12 Maggio 1605) ", *Opere Storiche*, Volume Secondo, 279-285, esp. 284-285 ; H. Bernard, *Matteo Ricci's Scientific Contribution to China*, 66-67. It is necessary to note here that the Chinese name of Father Clavius is " Ding ", the meaning of which is the same as that of the Latin " Clavis " original meaning " nail ".

According to the above description, we may guess that Li Zhizao had already had an opportunity to study Clavius's *Gnomonices Libri Octo* (Pé-táng, 1581 edition, n° 1301) with the assistance from Ricci in Peking. Undoubtedly, Li not only found those European scientific instruments fascinating, but also practical. His 1607 preface to the *Hungai tongxian tushuo* which was translated by himself, with the help from Ricci during the period 1605-1607, has this detailed account to say : " Confucian scholars are [associated] with practical learning (*shixue*)... In bygone years I acquainted with Father Ricci in Peking who is an European. He showed me the planispheric astrolabe (*pingyi*). The instrument in the form of a circular disc represents the great outer rim, from it different circles are drawn. Upper celestial sky and lower horizon, encircling fixed stars and heavenly bodies, the alidad with a hole attached on the back [of the instrument] seems like the heaven with canopy (*gaitian*). However, all the degrees must be read from the outer rim. Taking the centre as the North Pole, just tallies with what has been mentioned in [the Medical Classics] *Suwen* that north in the central and south in the outerbound. Well-arranged circles of equator, southern and northern tropic, in telling times and seasons, just tallies with the ultimate purpose of observing stars and the sun in Ancient China by two astronomical-officials Xi and He "[25].

On several occasions, Ricci described Li Zhizao's enthusiasm in translating culture. For instance :

(1) Ricci's letter to Father P. Girolamo Costo S.J. in Rome from Peking dated March 6, 1608, he writes : " He (i.e. Li Zhizao) learned mathematics with me. Now he returns to his hometown Hangzhou where it takes two months' journey from Peking. He has already compiled what he learned in a book. Last year he published *Hungai tongxian tushuo* which was a translation of Father Clavius's *Astrolabium*, and a document concerning way of use being attached. I have two to three copies at hand. Therefore I can only send one copy to Father General... From the illustrations and diagrams of the book, we are able to know Chinese people are talented enough to learn our Science and can achieve fruitful results "[26].

(2) Ricci's letter to Father Claudio Acquaviva the General, S.J. at Rome from Peking dated March 8, 1608. Ricci says : " I would like to send you a book which is translated by an eminent scholar Li Zhizao... He learned mathematics with me for a long time. This book is translated from Father Clavius's

25. *Hungai tongxian tushuo*, in *Tianxue chuhan* (First Collection of Writings on Learning from Heaven), composed by Li Zhizao in 1628 ; reprinted in Wu Hsiang-hsiang edited, *Chung-kuo shih-hsüeh ts'ung-shu* (Collectanea of Chinese Historical Studies) 23 (Taipei, Student Book Co., 1965), 1a-4a (1711-1717). On astrolabe, see R.T. Gunther, *The Astrolabes of the World* (2 vols, Oxford, reprint in one volume, 1976) and W. Hartner, " The Principle and use of the Astrolabe ", in *Willy Hartner : Oriens-Occidens*, Hildesheim, Georg Olms Verlagsbuchhandlung, 1968, 287-311.

26. M. Ricci, " Al P. Girolamo Costa S.I. a Roma (Pechino, 6 Marzo 1608) ", *Opere Storiche*, Volume Secondo, 330-339.

Astrolabium and entitled *Hungai tongxian tushuo*. Besides, *Description of the Heavenly Sphere (Tianti tu ?)* bas been published "[27].

(3) Ricci's letter to Father Claudio Acquaviva, the General, S.J. at Rome from Peking dated August 22, 1608. Ricci further points out that : " Li Zhizao has acquainted with me for nearly five years. He published my *Universal World Map* which was 15 feet in height and 18 feet in width. He learned mathematics with me for a long period. This year he published *Hungai tongxian tushuo,* it was a select translation of my teacher Father Clavius's *Astrolabium*. I teach him orally and he takes down with further embellishment... He is now returning to Peking to prepare in the publication of my teacher Father Clavius's *Epitome Arithmetricae Practicae* (1585 edition, Pé-táng n° 1296) and *De Horologiis [Horologiorum nova descriptio* (Romae, 1599, Pé-táng n° 1302) or *Compendium brevissimum describendorum Horologiorum Horizontalium ac Declinantium* (Romae, 1603 ; Pé-táng n° 1294)]. The latter has been translated into Chinese [but lost during the Ming-Qing transition]. He also made a number of beautiful and accurate astrolabes "[28].

LI ZHIZAO TRANSLATED WORKS REVISITED : WITH SPECIAL REFERENCE TO *YUANRONG JIAOYI* (1608), *TONGWEN SUANZHI TONGBIAN* (1613) AND *HUNGAI TONGXIAN TUSHUO*

In the third chapter " Rongjiaotu yi " of Ricci's main works *Qiankunti yi,* there are eighteen problems related to isoperimetric figures which have been discussed in Archimedes' *On the Sphere and Cylinder.* The *Siku quanshu* edition of *Qiankunti yi* only listed out Ricci's name as author. As we know, *Qiankunti yi* was a select translation of Clavius's commentary on Sacrobosco's *Sphere.* The third chapter " *Rongjiaotu yi* " is chiefly taken from Clavius's commentary *Caput primum* particularly from *caelum esse figurae sphaericae* to *de figuris isoperimetris, definitiones propositio I-XVIII*[29]. The question is that who assisted Matteo Ricci in translating this series of problems on isoperimetric figures into Chinese ?

On the other hand, Li Zhizao translated a mathematical treatise on isoperimetric figures into Chinese entitled *Yuanrong jiaoyi* under Ricci's direction in 1608, which was incorporated into *Tianxue chuhan* (First Collection of Writings on Learning from Heaven, 1628). Pasquale M. D'Elia and Li Yan (1892-

27. M. Ricci, " Al P. Claudio Acquaviva Prep. Gen, S.I. a Roma (Pechino, 8 Marzo 1608) ", *Opere Storiche,* Volume Secondo, 339-353.

28. M. Ricci, " Al P. Claudio Acquaviva Prep. Gen, S.I. a Roma (Pechino, 22 Agosto 1608) ", *Opere Storiche,* Volume Secondo, 353-369. But *Fonti Ricciane*'s record is slightly different, " Ha tradotta anco tutta la *Sfera [Hungai tangxian tushuo]* senza lasciar niente. Perchè, doppo che hebbe l'Euclide tradotto, voltò anco il *Trattato delle Figure isoperimetre [Yuanrong jiaoyi],* et anco quello delle *Costellationi delle stelle [Jingtian gai],* Con che fece il globo celeste [*tianqiu*], et anco il terrestre [*diqiu*], assai belli. " (177-178).

29. C. Clavius, *In Sphaeram Ioannis de Sacro Bosco Commentarius,* Romae, 1570, 102-140.

1963) concluded that *Yuanrong jiaoyi* was a translation based on Clavius's *Trattato delle Figure isoperimetre*[30]. Indeed, their observation was criticized by Itaru Imai who pinpointed that the translation was based on Clavius's *Sphaeram* instead[31]. And Jean-Claude Martzloff seems to follow Imai's view[32]. After careful comparison between Ricci's " *Rongjiaotu yi* " and Li Zhizao's *Yuanrong jiaoyi*, it is obvious that two pieces of translation are entirely the same and we may conclude that Li Zhizao was once helping Ricci to translate *Caput primum* of Clavius's *Sphaeram* into Chinese with a title *Rongjiaotu yi* in 1605.

In their effort to translate European fundamental mathematics, Ricci and Li Zhizao adapted Clavius's *Epitome Arithmaticae Practicae* into Chinese under the title *Tongwen suanzhi* (1613). In this book, western written computation (*bisuan*) and Venetian monk Honoratus's (1550 ?-1600 ?) " galley method " (known as *fanchuan fa* in Chinese) were introduced[33]. However, researchers mostly regarded *Tongwen suanzhi* simply as a translation of Clavius's *Epitome Arithmeticae Practicae* with the exceptions of Kusuo Takeda and Martzloff who maintained that the text was the result of a compilation of Clavius's text and numerous earlier Chinese arithmetics[34]. After an in-depth survey, I discover that the text also contains several adaptations from Clavius's *Geometricae Practicae* (Romae, 1604 Pé-táng n° 1300). For instance, *celiang sanlü fa* of chapter six in *Tongwen suanzhi tongbian* was a partial translation based on the section of *Earundem linearum rectarum dimensionem per Quadratum Geometricum exequens* in Liber III of *Geometriae Practicae*[35]. Illustrations of the construction of a quadrant (so-called *judu* in Chinese) and some surveying problems will be shown here.

30. *Fonti Ricciane*, p. 177, note 4 ; Li Yan, *Zhongguo suanxue shi* (*History of Chinese Mathematics*), Shanghai, Commercial Press, 1955 revised edition, 187.

31. I. Imai, " *Kenkon taigi zatsu kô* ", (Some Reflections on *Qiankunti yi*), in Yabuuchi Kiyoshi (eds), *Min Shin jidai no kagaku gijutsu shi* (*History of Sciences and Technology in Ming-Qing Period*), (Kyoto, Kyoto Daigaku Jinbun Kagaku Kenkyûsho, (1970), 35-47, esp. 45-46.

32. J.-C. Martzloff, *A History of Chinese Mathematics* (translated by S.S. Wilson), New York, Springer-Verlag, 1997, 22.

33. D.E. Sinith, *Rara Arithmetica*, Boston, London, Chelsea Publishing Co., 1908, 487. See also K.W. Fung, " Fang Zhongtong ji qi Shudu yan (Fang Zhongtong [1634-1698] and his *Shudu yan* [Development of Calculations and Measure]) ", *Lun Heng* (Hong Kong), 2 (1995), 123-204, esp. 183-185.

34. T. Kusuo, " Mindai sûgaku no tokushitsu : *Sampo tosô* seiritsu no katei (The Characteristics of Chinese Mathematics in the Ming Dynasty : the Elaboration of the *Suanfa tongzong*) ", *Kagakushi Kenkyu*, 28 (1954), 11-16 ; 29 (1954), 8-18. *Ibid.*, " *Dônbun Sanshi* no seiritsu (The Elaboration of the *Tongwen* suanzhi) ", *Kagakushi Kenkyu*, 30 (1954), 7-14. Martzloff, " The influence of Matteo Ricci's Mathematical works ", in *International Symposium on Chinese-Western Cultural Interchange in Commemoration of the 400th Anniversary of the Arrival of Matteo Ricci, S.J. in China*, Taipei, Fu Jen Catholic University Press, 1983, 439-453 ; *ibid.*, *A History of Chinese Mathernatics*, 22.

35. C. Clavius, *Geometriae Practicae*, Romae, 1604, Liber III, 64-103.

As I have already stated that the practical art of making an astrolabe attracted Li Zhizao to study and translate Clavius's *Astrolabium*. The 763 pages in this single-edition work provide a thorough geometrical and trigono-metrical basis for the theory of the instrument, and the construction of its geo-metrical projections. Since Li Zhizao realized the actual procedure to make an astrolabe, he therefore selected the main characteristics or concepts to trans-late. In two volumes of *Hungai tongxian tushuo*, Li Zhizao used 18 topics to explain the construction of an astrolabe. The 18 topics are as follows :

1. General Diagram and Explanation of an astrolabe's body (front and back ; for back, it shows the construction of the shadow divisions in qua-drants).

2. Diagram and Explanation of showing degrees of the outer rim i.e. Celes-tical circle (includes the south line/ the line meridional, the north line/ the line of midnight, the east line/ the line oriental, the west line/ the line occidental).

3. Diagram and Explanation of dividing equal hours according to the outer degrees.

4. Diagram and Explanation of setting up the terrestrial disc (includes azi-muth circle).

5. Diagram and Explanation of determining the zenith.

6. Diagram and Explanation of determining the horizon.

7. Diagram and Explanation of showing degrees of ascension (i.e. to find the position of the rising of the sun and of other planets).

8. Diagram and Explanation of finding the true position of stars.

9. Diagram and Explanation of diurnal and nocturnal arcs.

10. Diagram and Explanation of dividing twelve signs.

11. Diagram and Explanation of finding the time of the spring of the dawn and the end of the evening twilight.

12. Diagram and Explanation of setting up the ecliptic and the celestial disc.

13. Diagram and Explanation of Position of Fixed Stars (by using three kinds of calculation tables : a) ecliptic longitudes and declination ; b) right ascensions and polar latitudes ; c) ecliptic longitudes and latitudes).

14. Diagram and Explanation of setting up 24 seasonal periods with Chinese celestial circle (i.e. $365 1/4°$)

15. Diagram and Explanation of showing the shadows of a gnomon at tra-ditional Chinese six temporal divisions.

16. Diagram and Explanation of finding the height of things or positions by point of shadows according to theory of right-angled triangle.

17. Diagram and Explanation of alidad's degree divisions (includes flat-ruler type [*fencheng chi shi*] and type with holes to measure the altitude of the sun [*kui tong shi*]).

18. Diagram and Explanation showing the way to use an astrolabe.

Most of the illustrations or diagrams in the *Astrolabium* had been modified by Li Zhizao and his accurate drawings earned Matteo Ricci's high appreciation. As far as I know, Li Zhizao's contribution in translating Clavius's *Astrolabium* had great influence upon his contemporaries. Ming loyalist like Xiong Mingyu (1579-1649) and Fang Kongzhao (1591-1655) had cited illustrations or passages from *Hungai tongxian tushuo* in their works[36].

To conclude, Clavius's mathematical and astronomical textbooks represented Jesuit orthodoxy on scientific matters. His viewpoints were transmitted to East Asia by his faithful student Matteo Ricci and became more widespread after some of his works were translated into Chinese. As a collaborator of Ricci, Li Zhizao also published books related to Western Sciences. His *Tianxue chuhan* does not only include Clavius's works, but also contains other Western works. It exerted great influence upon intellectual circles in China, Japan and Korea[37].

36. K.W. Fung, " Mingmo Xiong Mingyu *Gezhi cao* neirong tanxi (A Critical Study of Xiong Mingyu's (1579-1649) *Gezhi cao)* ", forthcoming in *Ziran kexueshi yanjiu (Studies in the History of Natural Sciences,* Beijing), 16 (1997), 304- 328.

37. K.W. Fung, " Fang Yizhi yu Sanpu Meiyuan (Fang Yizhi and Miura Baien) ", *Journal of Oriental Studies* (Hong Kong), 33 (1995), 1-31.

Bisuan 筆算
celiang sanlü fa 測量三律法
fanchuan fa 帆船法
Fang Kongzhao 方孔炤
fengcheng chishi 分成尺式
Gaitian 蓋天
He 和
Hungai tongxian tushuo 渾蓋通憲圖説
Hunxiang tu 渾象圖
Itaru Imai 今井湊
jinshi 進士
Jiuchongtian tu 九重天圖
judu 矩度
kui tong shi (目+規) 箶式
Li Wocun 李我存
Li Yan 李儼
Li Zhizao 李之藻
pingyi 平儀
Qiankunti tu 乾坤體圖
Rongjiaotu yi 容較圖儀
Shanhai yu di quantu 山海輿地全圖
shixue 實學
tiandi yi 天地儀
Tianti tu 天體圖
Tianxia zongtu 天下總圖
Tongwen suanzhi tongbian 同文算指通編
Suwen 素問
Xi 羲
Yuanrong jiaoyi 圜容較儀

THE EARLIEST EVIDENCE OF THE INTRODUCTION OF KEPLER'S LAWS INTO CHINA AS IS OBSERVED IN THE *LIFA WENDA*

Keizo HASHIMOTO

Recently, Catherine Jami and I have found the manuscript of the *Lifa wenda* (*Dialogue on Astronomy*) by Jean-François Foucquet (Fu Shengze, 1665-1741) at British Library[1]. Together with the other, partial, but otherwise identical version, which she had located at the Vatican Apostolic Library[2], the manuscript, especially Book v, Part 1, from the British Library, gives us the detail of how Kepler's first and second Laws introduced into China as early as in the 1710s. This means that we can go back the history of the introduction more than two decades earlier than the so far believed date.

On the other hand, when the *Chongzhen lishu* had been compiled for several years from 1629, although Kepler's optical astronomy was extensively introduced, we cannot find the slightest evidence of the description of his Laws in it. Thus, the *Lifa wenda* can be regarded the earliest evidence, in which Kepler's Laws were openly discussed.

In connection with Kepler's Laws, Copernicanism was also discussed particularly concerning the instrumental model of the solar system, that is the Orrery, manufactured by O.C. Roemer, which had been brought to China by the French Jesuit missionaries and presented to Kangxi Emperor as from the French King, Louis XIV[3]. On several pages of the manuscript, Copernicanism is referred to in connection with the explanation of planetary motions as well. The latter of the evidences shows that Copernicanism had been introduced into China almost half a century before M. Benoist did in his *Treatise on the Earth* (*Kunyu quantushuo*) in 1767[4]. Both of them can drastically change our under-

---

1. Oriental and India Office Collections, OR Add. 16634.

2. Borgia Cinese 319(1) and 319(2). *Cf.* K. Hashimoto, C. Jami, " Kepler's Laws in China : A Missing Link ? ", *Historia Scientiarum,* vol. 6-3 (1997), 171-185.

3. A. Nissen, *Ole Roemer,* Copenhagen, 1944, 32.

4. K. Yabuuti, *Chugoku no tenmon rekiho,* Tokyo, Heibonsha, 1969, 171.

standing of the history of astronomy in China. First we should like to discuss the problem of the early introduction of Kepler's Laws.

KEPLER'S LAWS

First of all, let us see how Kepler's Laws were accepted and developed in Europe after Kepler's introduction of elliptic orbits, replacing the combination of circular motions. According to Curtis Wilson, Ismael Boulliau was objecting to the magnetic mechanism hypothesised by Kepler to account for the eccentricity of the planetary orbits[5] and he composed the *Astronomia philolaica* in 1645. Boulliau had imagined another way of deriving elliptical orbits from uniform circular motions, proposing for the first time, " from the general circumstances of planetary motion ", that these orbits are elliptic. According to Boulliau, the circles lie in a cone and the mean motion takes place about the axis parallel to the base[6]. He failed to recognise that this implies an equivalent uniform angular motion about the non-solar focus; the hypothesis is thus equivalent to the empty-focus equant that Kepler's " Uranian friend " Albert Curz had proposed for the Moon (to which Kepler refers in the *Rudolphine Tables,* 1627).

In the *Astronomia geometrica* of 1656, Seth Ward assumes that the " simple elliptical hypothesis " with superior focus as equant point is true. Both Kepler and Boulliau had failed to recognise its truth. Another proponent of the simple elliptical hypothesis was Emile-François Pagan, who in 1657 published *La théorie des planetes*[7].

Now let us see how Foucquet describes and introduces the discovery and development of the theory of elliptic orbits in the *Lifa wenda*, which must have been prepared probably between 1712 and 1716[8]. In Book v Part 1, he begins his discussion with repeatedly alluding to the recent sixty-years development of the astronomical instruments and observational achievements in Europe[9], particularly after the establishment of the Paris Observatory and the installation of the telescope mounted with micrometer (*liang wei ge*[10]) there.

To begin with, Foucquet discusses the necessity of introducing the elliptical orbits in place of the combination of circular motions. Before the explanation

5. C. Wilson, " Predictive astronomy in the century after Kepler ", R. Taton, C. Wilson (eds), *Planetary Astronomy from the Renaissance to the Rise of Astrophysics.* Part A : *Tycho Brahe to Newton. The General History of astronomy,* vol. 2A, Cambridge University Press, 1989, 161-206, 172.

6. C. Wilson, 1989, 173.

7. The full title is *La theorie des planetes du Comte de Pagan, ou tous les orbes celestes sont geometriquement ordonnez, convert le sentiment des astronomes. Cf.* C. Wilson, 1989, 178.

8. ARSI, Jap. Sin. II, 154.

9. *Lifa wenda,* V-1-1. In the introduction of the Treatise on Lunar Motion, Foucquet first discusses this topic in detail in Chapter III-1.

10. Book III, Part 1, ff. 72a-b.

of the problem of Martian motion, he particularly emphasises the shift of the perihelion of the orbit of Mercury in order to demonstrate the inadequacy of circular motions. Then he describes how the recent observational results show the discrepancy from the theoretical calculations.

Foucquet tried to show how recent telescopic observations had become precise by reporting Huygens's determination of Saturn's ring making use of his long telescopes from 3rd March 1655, through 16th 1656, to 12th 1659[11]. In the manuscript *Lifa wenda,* he explains the result with the heliocentric model of the solar system as Huygens did. Foucquet also tries to emphasise the importance of Cassini's telescopic observations of the surface of Jupiter, including a dark spot appeared between 1690 and 1691[12].

Kepler had derived the elliptic orbit of Mars making use of Tycho's observations of opposition. We can observe four observational data reduced from Tycho's observations in the *Lifa wenda*[13]. We can find this data cited from the *Almagestum novum* in 1651 by the Bolognian Jesuit astronomer, G.B. Riccioli, of which had been made use by Boulliau. Based on these data, together with other observational results by various astronomers in Europe, Foucquet tries to emphasise the inevitability of the introduction of the non-circular motions of the planets.

Then he starts to discuss the mechanistic " necessity " of the introduction of elliptic orbits, alluding to the Cartesian physics, which we can observe immediately after the discussion of the non-circular motion of Mars[14]. We shall see this problem below soon.

Although Foucquet failed to give any illustrations to explain the geometrical orbit of planets in the manuscript, we can reconstruct what he means to describe, that is, the elliptic orbits in terms of Kepler's method. Following the explanation of Kepler's Laws, he discusses Boulliau's so-called revised method, Pagan's (Bagang) simple method, and Riccioli's (Lizhuoli) spiral orbit, successively.

Let us see his description of so-called Kepler's method (*Keboer zhi fa*), in place of the areas rule, in the manuscript itself. He writes as follows : after having studied the record of observations of Mars by Tycho Brahe, " Kepler for the first time abandoned circular motions, and adopted the ellipse (*Danxing-xian*) for the orbit of Mars "[15].

11. V-1, ff. 3a-5b. The micrometer is described here.

12. V-1, 8b. *Cf.* The dark spot on Jupiter produced by the impact of Shoemaker-Levy 9 comet in 1995. As to Cassini's observation, see I. Tabe *et al.*, " Discovery of a Possible Impact Spot on Jupiter Recorded in 1690 ", *Pub. Astron. Soc. Japan,* 49 (1997), p. L1-L5.

13. V-1, ff. 36a-45b.

14. V-1, ff. 43a-45b.

15. V-1, f. 47b.

Here he did not use the term *Tuoyuan*[16], which became the standard representation after the compilation of the *Shuli jingyun* in 1723. In order to explain the new method, he describes the geometry of ellipse, emphasising the importance of understanding the character of it.

Foucquet constructs the ellipse, on which a planet moves, and also its auxiliary circle, and discusses the properties of the auxiliary circle. Firstly, he draws the auxiliary line, which is perpendicular to the major axis, then he shows that the ratio HF:GF is proportional to the ratio of the minor axis of the ellipse to the diameter of the auxiliary circle. Here HF is named the *liexian* and GH the *zhengxian*. And he goes on to show that the ellipse has two focuses (*juguang juhuodian*, or abridged as *judian*), on the lower of which the sun is located.

After that Foucquet explains the areas rule of Kepler[17]. He discusses how the planet moves on the ellipse about the sun. And, for the explanation, he makes use of Riccioli's representation, which we can read in the *Almagestum novum*. Riccioli tries to discuss the areas rule introduced by J. Kepler[18]. Obviously Foucquet must have been describing the method, as a whole, relying on the contents of explanation as well as the illustrations from Riccioli's book.

He also alludes to the cause of the elliptical motion of planets like magnetic force, which Kepler used. But, he declines Kepler's analogy, and follows Descartes' physics of the cause of motion. By doing so, he starts to introduce Boulliau's method as well as Pagan's so called simple method, which used the second focus as the equant.

It is interesting to observe Cassini's method explained as the fourth method in the *Lifa wenda* under discussion. He assumed the simple elliptic hypothesis ; if S is the focus where the eye (or Sun or Earth) O is the centre of the ellipse, *ARP* a circle with radius equal to the semi-major axis of the ellipse, and if the true anomaly is $v_R = <ASR$ and the mean anomaly M_R, then $<AOR=1/2$ (M_R+v_R), where $AB = v_A - v_B$, $DF = M_A - M_B$, $BC = v_B - v_C$, $DF = M_B - M_C$: only the differences in true and mean anomaly being observationally determinable[19]. The intersections G and H fix the line GH on which point O, the centre of the ellipse, is found by dropping a perpendicular from B. As the text suggests[20], he invented *cassinoid*, with the aim of obtaining a possible orbit for

---

16. In the preface to the treatise on planetary motions, we observe as the more term, *tuoyuan*, has been used for the shape of orbits, oval or ellipse, other than circle (v-1, f. i). The term, *tuoyuanenxing*, with the hand radical for the character, *tuo*, has first appeared in the *Celiang quanyi, quan* 6, in the *Chongzhen lishu*, where the conic sections are discussed. See the *Xinfa siianshu* edition, *quan* 92, p. 9a, 1.2 ; Taibei reprint version, 1972.

17. v-1, ff. 50a.

18. v-1, ff. 51b-52b.

19. *Cf.* C. Wilson, 1989, 182-183.

20. v-1, ff. 78a-83b.

the planets in which the superior focus would serve as equant point, and introduced another kind of elliptic orbit[21].

As the fifth method he also explains Riccioli's spiral orbit of the sun (or planets). We just would like to show this from the figure from Riccioli's original book. As a whole, we can repeatedly say that he is rather faithfully following the *Almagestum novum* by G.B. Riccioli.

However we must point out that, the long discussion of Kepler's Laws notwithstanding, Foucquet eventually transcribed as the astronomical tables La Hire's *Tables* in 1702, instead. La Hire had produced the *Tables,* totally relying on his own observations made for long time at the Paris Observatory after he succeeded Jean Picard. It was, indeed, practical astronomy, which Foucquet was trying to introduce in the manuscript, although he tried to explain even the physics of planetary motions.

In this connection it is also worthwhile to point out that he brought to China the achievements of determination of astronomical constants which were remarkably improved after the invention of the telescopes mounted with micrometer. To mention a few examples : the obliquity of ecliptic, Cassini's determination of horizontal parallax of the Sun as well as the refraction up to the zenith. Chinese astronomers had only known Tycho's refraction, which had introduced several decades ago.

COPERNICANISM

In the *Almagestum Novum* of 1651, G. Riccioli rejects the ellipse because he does not find the empirical evidence for it sufficiently convincing. But, in his later book, the *Astronomia reformata* of 1665, Riccioli is anti-Copernican as ever, claiming in fact to have proved the immobility of the earth. However, he has then adopted the ellipse as provisional basis for planetary theory[22]. Nevertheless, he discussed the theory of Copernicus (Gebaini) as the hypothesis of the universe when introducing the semi-Tychonic world system[23].

It seems likely that the Jesuit missionaries in China, who took care of the Imperial Astronomical Observatory in Beijing, eventually followed Riccioli's choice for the adoption of Kepler's Laws. It was crucial for them to compile the *Lixiang kaocheng houbian* (*Sequel to Astronomical Compendium*) in 1742. This is until now believed to be the first, formal introduction of Kepler's theory of elliptic orbits, for which the German Jesuit missionary, Ignatius Koegler (Dai Jinxian), together with the Portuguese missionary, Andreas Pereira, took

21. *Cf.* C. Wilson, 1989, 183.

22. C. Wilson, 1970, 111.

23. III-1, f. 6b. As to the " semi-Tychonic " world system of Riccioli, see Ch. Jones Schofield, *Tychonic and Semi-Tychonic World Systems in the Seventeenth Century,* New York, Arno Press, 1981, and " Tychonic and semi-Tychonic World Systems, R. Taton, C. Wilson, *ibid.,* 32-44, esp. 40.

responsibility to the compilation as the Astronomer Imperial (Qintian jian-zheng) of the Qing dynasty.

In connection with the problem of Copernicanism, or heliocentrism, we should not overlook Foucquet's discussion of the method of the determination of the longitude, making use of the satellites of Jupiter, which Galileo had originally suggested. In the discussion of the determination of differences in terrestrial longitude, the discovery of finite velocity of light by O.C. Roemer, making use of the satellites of Jupiter, played an important role[24].

But, here, we must point out that Roemer's planetary model, based on the heliocentric idea, has been explained in detail as the appendix to the Treatise on Eclipses in the *Lifa wenda*[25]. The instrument was presented to Kangxi Emperor from Louis XIV, when the French Jesuit missionaries arrived in Beijing[26].

Now, we have to continue to discuss the problem of the determination of geographical positions. Cassini's instructions furnished a clear picture of the best seventeenth century research methods and at the same time explained how terrestrial longitude could be determined by timing the eclipses of the satellites of Jupiter[27]. The most satisfactory time observations of Jupiter could be made of the immersions and emersions of the first satellite.

When Jean de Fontaney, a Jesuit professor of mathematics at the College of Louis le Grand, was preparing to leave for China as the head of the first French Jesuit mission, J.D. Cassini trained him. On his way to China he contributed data on the longitudes of the Orient[28]. Later the French missionaries worked for the imperial enterprise of the survey for the map of China in the late 1710s.

Foucquet, depending on the *Traité de la lumière* of 1690 by C. Huygens, discussed this method of geodesy. We can see the explanation of the method, which faithfully reflects Huygens's discourse[29]. And Foucquet categorically asserts that " the satellites of Jupiter were best made of for cartography by making use of the method, which relied on the velocity (*liuxing*) of light, which had been determined by Roemer "[30].

24. S. Debarbat, C. Wilson, " The Galilean satellites of Jupiter from Galileo to Cassini, Roemer and Bradley ", R. Taton, C. Wilson (eds), *op. cit.*, 1989, 144-157, p. 156.

25. III-1, paragraph 14, which has been omitted from the British Library version, the part of which, otherwise, seems to be copied from the original version. The omission means that the censorship seemed to be executed.

26. A. Nissen, *Ole Roemer,* Copenhagen, 1944, 32. Later Roemer manufactured the model based on the Tychonic world system in Paris.

27. L.A. Brown, *The Story of maps,* 1949, reprint ed., New York, Dover, 1979, 221ff.

28. *Ibid.,* 220.

29. V-2, f. 33a. Fig. from Engl. trans. by S.P. Thompson, *Treatise on Light,* 1912, Dover ed., 1962, 8.

30. V-2, f. 32b.

It is interesting enough to read in the beginning of the discussion that Foucquet meant to draw the illustration of the model to show the motion of the first satellite of Jupiter. He does not try to revise Roemer's model of the solar system. Instead, he faithfully explains the model. For this purpose, he says that the sun " stands still " (budong)[31] without any motions in the centre of the orbit of Jupiter as well as of that of the Earth.

THE CARTESIAN BASIS OF FOUCQUET'S PHYSICS

Lastly we should like to see Foucquet's scientific knowledge of physics as it is reflected in the manuscript on which we are discussing. This is crucial in order to understand how Foucquet recognised the cause of planetary motions. Here we can say that his discussion was clearly based on Cartesian mechanism.

As for the mechanism of planetary motions, he first explains Kepler's magnetic mechanism as the cause of planetary motions. Then he moves on to introduce Boulliau's ellipse, because he followed Riccioli, who had not agreed with the analogy of magnetism used by Kepler.

Foucquet assumes that the Sun is located in the centre of the great circle of planetary orbits, and that it is the essence of the fire (huo zhi jing) and can be the source of the motions (dong zhi yuan)[32]. The five planets all receive its movement, because they are in the aether which transmits the motion from the Sun. It is the moving power of the aether (jingqi zhi nengli). And Foucquet gives to discourse the reason why the power causes the non-circular motions of planets[33]. It is because, as he tells us, Descartes stated that the motions in the heaven were not perfectly circular[34]. First Foucquet tries to show that, if the centrifugal force works, it should cause the circular motions of planets. In order to explain the mechanism Foucquet cites Descartes' second law of nature from the *Principles of Philosophy* and explains the planetary motion, although the figure is missing from the manuscript, depending on the Cartesian illustrations.

He describes that in the heavens the aether exists and expels the five planets, and says that if the expelling power of aether (jingqi) comes from the sun straight and uniform then the revolution of the planets should be circular. But, as the manuscript says, it obliquely works, so that the orbits of the planets should be elliptic[35]. We must understand that his suggestion came from Leibniz's theory of planetary motion. Leibniz supposed that " the vortex carrying

31. v-2, f. 33a.
32. v-1, f. 45a.
33. v-1, ff. 43a-45b.
34. R. Descartes, *Principia Philosophiae*, III-34, 1644.
35. v-1, f. 45b.

the planets rotated in spherical layers and in order to account for the elliptical orbit, he postulated two motions of the planets ; a radial motion from layer to layer and a trans-radial motion in which the planet moved with the same speed as the fluid "[36].

In a few words, we can repeatedly say that Foucquet's concern with the cause of the planetary motions centred on Cartesian physics.

Then, as to the propagation of rays in straight line, he continues to discuss the problem in terms of the concept of the medium of subtle aether (*jingqi*), relying on the physical arguments by Descartes as well as Huygens[37]. It is interesting enough to find that he here mentions the idea of vortex (*xuanquan zhi ho bo*) in the medium[38]. Here it is interesting to read that he transcribes " physics " into the Chinese term, *gewu qiongli zhi xue*[39].

Lastly, as to the propagation of the light, the manuscript discusses that the luminous body impulses the surrounding aether, and propagates the ray in all directions. We have round the whole discussion here has been comprised of the complete translation of the *Treatise on Light* by Christiaan Huygens in 1690, including the discussion of Roemer's determination of the velocity of light and that of the longitude, using the first satellite of Jupiter[40]. We can find this extensive discussion in the chapter of Part 2, Book V, of the *Lifa wenda,* where Foucquet discusses the problem of determination of longitude by making use of it[41]. We can definitely say that Huygens was a crucial authority for Foucquet in the explanation of this optical problem as well.

The next step we must take is the more detail analysis of the nature of the treatise on planetary motions, that is Book V of the manuscript and the problems of the solar and lunar motions together with the treatise on eclipses which consists of the first three books of it. And, our future work should be carried out concerning the historical fact of whether or not Foucquet's introduction of Kepler's Laws had any significant effect on the compilation of the *Lixiang kaozheng houbian* in 1742.

36. E.J. Aiton, " The vortex theory in competition with Newtonian celestial dynamics ", in R. Taton, C. Wilson (eds), *Planetary astronomy from the Renaissance to the rise of astrophysics,* Part B : *The eighteenth and nineteenth centuries. The General History of Astronomy, 2B,* Cambridge UP., 1989, 3-2l, especially 10.

37. v-2, f. 43a.

38. v-2, ff. 44a-45a.

39. v-2, f. 32b.

40. *Treatise on Light*, Engl. trans. by S.P. Thompson, New York, Dover, 1962.

41. v-2, ff. 33a-46a.

Bagang 巴岡
budong 不動
Celiang quanyi 測量全義
Chongzhen lishu 崇禎曆書
Dai Jinxian 戴進賢
Danxing xian 蛋形線
dong zhi yuan 動之原
Fu Shengze 傅聖澤
Copernicus, Gebaini 歌白尼
gewu qiongli zhi xue 格物窮理之學
huozhijing 火之精
jingqi 精氣
jingqi zhi nengli 精氣之能力
judian 聚點
juguang juhuodian 聚光聚火點
Kebo'er zhi fa 刻白爾之法
Kunyu quantu shuo 坤輿全圖說
Lifa wenda 曆法問答
Lixiang kaocheng houbian 曆象考成後編
Lizhuoli 利酌理
liangwei ge 量微格
liuxing 流行
Qintian jianzheng 欽天監正
Shuli jingyun 數理精蘊
Tuoyuan 橢圓
xinfa suanshu 新法算書
xuanquan zhi bo 旋圈之波
zhengxian 正弦

GUIMAO YUAN CALENDAR (1742-1911) AND ISAAC NEWTON'S THEORY OF THE MOON'S MOTION[1]

Lu Dalong

Lixiang kaocheng houbian (1742, *Sequel to the Compendium of Observational and Computational Astronomy*, and one of the Chinese astronomical theories and Chinese calendars) was derived from Isaac Newton's theory of the Moon's motion.

In the first half of the eighteenth century, Isaac Newton's original masterworks, especially the magisterial *Philosophiae Naturalis Principia Mathematica* (1726, the 3rd Latin edition), together with contemporary commentaries, among which were David Gregory's *Astronomiae Physicae et Geometricae Elementa* (1726, the 2nd Latin edition), William Whiston's *Praelectiones Astronomicae* (1707, the 1st Latin edition) and two volumes of John Harris' *Lexicon Technicum* (the first in the 3rd edition of 1716, the latter in the 1st of 1710), were brought into China, probably by missionaries.

From 1736 to 1742, in order to compile a new Chinese calendar, Roman Catholic missionaries Ignatius Koegler (1680-1746) and André Pereira (1692-1743) had cooperated with many Chinese astronomers, and resulted in *Lixiang kaocheng houbian.*

Guimao yuan Calendar, as an appendix to *Lixiang kaocheng houbian,* had been put into use in China from 1742 to 1911. It is possible that it was referred to the astronomical solar tables and lunar tables in William Whiston's *Praelectiones Astronomicae.* More importantly, Ignatius Koegler (1680-1746) and André Pereira had directly introduced Newton's basic concepts, data and calculus methods, which were relative to the several inequalities and perturbations of the Moon's motion in Newton's *Principia* (1713), into *Lixiang kaocheng*

1. The author gratefully acknowledges the support of K.C. Wong Education Foundation, Hong Kong. Meantime, the author personally acknowledges the guidance of Professor D.T. Whiteside and the advice of Dr. Nick Kollerstrom.

houbian, without distinct expression of Newton's theory of universal gravitation which was in contradiction to missionary's doctrine.

SYNOPSIS OF THE CHINESE CALENDARS IN THE QING DYNASTY (1644-1911)

In the Ming Dynasty (1368-1644), Huihui Calendar (1383-1644) and Datong Calendar [1385-1644, in fact, it was Shoushi Calendar (1281-1368) of the Yuan Dynasty (1271-1368)], of corresponding reference, had been put into use in China. Because of that the Empire Astronomy Bureau was unable to forecast the solar eclipse which happened on 15 December 1610, some Empire astronomers put forward the suggestion of translating and introducing the Western astronomical theories, which was not to become true[2]. The Empire Astronomy Bureau had not forecast the solar eclipse once again on 21 June 1692. However, having made use of the Western astronomical theory, Xu Guangqi (1562-1633) calculated it with comparative accuracy. In the same year, Xu Guangqi, supported by the Emperor and having the aid of Catholic missionaries, was appointed to be in charge of revising the Chinese calendar by translating some Western astronomical books, and compiled *Chongzhen lishu* in 1634, which was not put into use because of the fall of the Ming Dynasty.

In the beginning of the Qing dynasty (1644-1911), German missionary John Adam Schall von Bell (1592-1666) revised *Chongzhen lishu* and dedicated it to the Emperor Shunzhi of the Qing dynasty (1644-1662) in the new name of *Xiyang xinfa lishu.* As to the prediction of solar eclipses, *Xiyang xinfa lishu* was more accurate than the old ones, and promulgated by the Emperor in the name of Shixian Calendar (1645-1741)[3].

Up to 1714, because of the serious defects of the ambiguities of explanations, and the inconsistency between the figures and tables in Shixian Calendar, the Administration of the Empire ordered the Empire Astronomy Bureau to revise *Xiyang xinfa lishu.* The revising work had been accomplished in 1722, and resulted in the publication of *Lixiang kaocheng,* the *Compendium of Observational and Computational Astronomy),* which is called Jiazi yuan Calendar as it selected the year of 1684 as its epoch.

Just as the historians of Chinese astronomy pointed out the astronomical theory upon which *Lixiang kaocheng* was based fell behind[4]. The prediction of the solar eclipse on 15 July 1730 was not inconsistant with its observation. Therefore, Ming Antu (*c.* 1692-1763), the minister of the Empire Astronomy

2. D. Shiran *et al.* (eds), *Zhongguo kexue jishu shigao,* vol. 2, Beijing, Kexue chuban she, 1982, 198-199.

3. C. Zungui, *Zhongguo tianwen xue shi,* vol. 3, Shanghai, Shanghai renmin chuban she, 1984, 1489-1492.

4. Zhongguo tianwen xue shi zhengli yanjiu xiazu bianzhu, *Zhongguo tianwen xue shi,* Beijing, Kexue chuban she, 1987, 232-233.

Bureau, presented a memorial to the Emperor and got his permission of invit-
ing the Bureau's missionaries Ignatius Koegier (1680-1746) and André Pereira
(1692-1743) to be in charge of the revision, which result was a calendar table.
Koegler and Pereira did not explain the theory upon which the calendar table
was based, and reveal the method of using the table, which was directly
arranged as an appendix to *Lixiang kaocheng*. In the Bureau, it was only Ming
Antu who was able to refer to this table, and it is certain that the Administration of the Empire was not satisfied with this result.

In 1736, replying to the memorial of Gu Cong, the minister of the Ministry
of Official Personnel, the administration of the Qing dynasty organized a
group, including Koegler, Pereira, Ming Antu, Mei Juecheng, He Guozong,
and many other astronomers in and out the Bureau, to revise and enlarge the
illustration of the tables and the explanation of the figures. All of this work was
completed in 1742 and resulted in the compilation of *Lixiang kaocheng houbian* in 10 volumes, in which the year of 1723 was selected as the epoch of its
calendar table, which is historically called Guimao yuan Calendar. Guimao
yuan Calendar had been in practice in China from 1742 to 1911, of long duration of 170 years. Without question, it is important to reveal the astronomical
theory from which Guimao yuan Calendar was derived.

THE THEORY OF THE MOON'S MOTION IN GUIMAO YUAN CALENDAR

The astronomical theory, which is called *lili* in China, is the theoretical
foundation of an astronomical calendar. The astronomical theory of Guimao
yuan Calendar is Isaac Newton's theory of the Moon's motion [the Chinese
name for Isaac Newton (1642-1727) was Nai Duan at early stage], the distinct
evidence for this conclusion is provided by *Yueli shuli* (the mathematical and
physical theory of the lunar equations), which is the volume 2 of *Lixiang
kaocheng houbian,* and composed of 11 sections, such as *Yueli zonglun* (General introduction on the lunar equations).

From a paragraph in *Yueli zonglun*, we can get a hint of the relation between
Lixiang kaocheng houbian and Newton's theory of the Moon's motion :
" From the time when Johann Kepler (Ke Bai'er, 1571-1630) invented the
method of Ellipsis, it has been granted that the orbits of the celestial bodies are
eccentric. The increase and decrease of the distance between two centres of the
orbit, and also of the velocity of the Moon, are dependent upon the orbit and
velocity of the Sun. Because of the change of the velocity of the Sun, the mean
motions of the Moon, the apogee and the node are changing, which inequalities is called First Average (*yi pingjun*). Because of the change of the distance
from the aphelion of the Moon to the Sun, there are the direct and retrograde
motions of the centre of the Moon's orbit, the change of the distance between
the two centres, and the inequality of the areas described by the mean motion
of the Moon, which is called Second Average (*er pingjun*). The maximum of

Second Average is called Supreme Average (*zui gaojun*). Because the Arctic Pole moves round the pole of the ecliptic, there is an advance or retrograde motion of the Moon's path, which results in the inequality of the Moon as it moves at the path, such inequality is called Third Average (*san pingjun*). In the past, there are no such four inequalities, which have been originated with Newton and others with time and again observations from the time of Kepler "[5].

Each of four concepts above mentioned has its astronomical meaning. In the sections, such as " It is required to find First Average ", " It is required to find Second Average " and " It is required to find Third Average ", of *Yueli shuli*, there are not only the discussion of the physical causes, but also the calculate of theoretical data for these concepts. Essentially, there is a consistance between these data and that of Newton's theory of the Moon's motion which is described in his magisterial *Philosophiae Naturalis Principia Mathematica* (1726, in the abbreviation form as *Principia*)[6]. Now, the definitions of First Average, Second Average, Third Average and Node Average (*jiaojun*, just as " the equation of the mean motion of the nodes "), which are given correspondingly in *Lixiang kaocheng houbian* and *Principia*, are juxtaposed as follows.

First Average

In the section " It is required to find First Average " of *Yueli shuli*, it is said that : " From the time of Kepler, Newton and the others have observed and tested again and again. It is pointed out that there is no Primary Average (*chujun*) when the Moon moves in apogee (*yuandi dian*) and perigee (*jindi dian*). The mean motion of the Moon (*taiyin pingjun*) is always slow, the mean motion of the apogee (*zuigao pingjun*) and mean motion of the node (*zhengjiao pingxing*) are always fast when the Sun moves behind perihelion (*jinri dian*) of the Earth. The mean motion of the Moon is always fast, the mean motion of the apogee and of the node are always slow when the Sun moves behind aphelion (*yuanri dian*). Therefore, in the mean distance of the Sun from the Earth, the annual equation of the mean motion of the Moon rises to 11'50", the greatest equation of the mean motion of the apogee comes out 19'56", and the greatest equation of the mean motion of the nodes 9'30". The annual equations of these motions, which are as the greatest equation of the Sun's centre to the equation of the Sun's centre, is named First Average "[7].

In *Lixiang kaocheng houbian*, while the greatest equation of the Sun's centre (*taiyang zhongju junshu*) given as 1°.56'13", the annual equation of the mean motion of the Moon (*taiyin yi pingjun*) is calculated as 11'49", the greatest equation of the mean motion of the apogee (*zuigao yi pingjun*) 19'56", and

5. *Yueli shuli*, " Yueli zonglun ", *Lixiang kaocheng houbian*, 2, Lizhi shuwu keben, 1896.

6. I. Newton, *Philosophiae Naturalis Principia Mathematica*, London, J. Streater (2 line, first issue), 1687 ; Cambridge, Cornelius Crownfield, 1713 ; London, Guil. & John Innys, 1726.

7. *Yueli shuli*, " Qiu yi pingjun ", in *ibid*.

the greatest equation of the mean motion of the nodes (*zhengjiao yi pingjun*) 9'29".

In Newton's *Principia,* there are some corresponding data, just as *Motus autem solis est in duplicata ratione distantiae terrae a sole inverse, & maxima centri aequatio, quam haec inaequalitas generat, est 1gr.56'.20" praedictae solis eccentricitati 16 11/12 congruens* [460/658, i.e., page 460 in *Principia* (1726) and page 658 in *Principia* (1972)][8], *& aequatio annua,... in mediocri solis a terra distantia ad 11'50"*, and *Unde prodit aequatio maxima medii motus apogaei 19'.43", & aequatio maxima medii motus nodorum 9'.24"* (460/659). The reason for selecting different values for First Average in *Lixiang kaocheng houbian* and *Principia*, is determined by the different values of the greatest equation of the Sun's centre, which the calculating methods are one and the same in *Lixiang kaocheng houbian* and *Principia*.

Second Average

In the section of " It is required to find Second Average " of *Yueli shuli*, it is said that : " It has been said that there is no Primary Average when the Moon moves in apogee and perigee of its orbit, and there is the annual equation when the Sun moves before and behind aphelion and perihelion of its orbit. Otherwise, there is no the annual equation if the Sun is in the positions of aphelion and perihelion. From the time of Newton, the concise observations have been repeatedly taken... Therefore, the maximum value for the another equation of the Moon's mean motion comes to 3'34" when the Sun is in the position of its aphelion, the Moon' apogee is in the octants of the Sun and the Moon in the middle of the path from the mean distance between the Sun and the Moon to the perigee (*yue tian gaobi zhongju hou sishiwu du*), and it rises to 3'56" when the Sun is in perihelion ; and it is subtracted from the mean motion while the Moon moves behind apogee and perigee of the Moon's orbit, and it is added to the mean motion while the Moon moves behind the mean distance between the Sun and the Moon. When the Moon's apogee is without the octants, the one part of the equation, which is related to the distance between the Sun and the Moon's apogee, is as the radius to the sine of double the distance of the Moon's apogee from the Sun ; and the other part, which is related to the distance between the Sun and the Earth, is as the cube of the distances of the aphelion and perihelion of the Sun from the Earth. So the equation, which is proportional to the cube of the distance of the current Sun (*benri taiyang*) from the Earth, is named Second Average "[9].

---

8. I. Newton, *Philosophiae Naturalis Principia Mathematica, the third edition with variant readings,* 2 vols., assembled and edited by A. Koyre and I.B. Cohen, Cambridge, University Press, 1972.

9. *Yueli shuli,* " Qiu er pingjun ", in *ibid.*

In *Principia*, Newton distinctly expresses that, by the theory of gravity (*per theoriam gravitatis*), he likewise has found another equation of the Moon's mean motion, depending upon the situation of the Moon's apogee in respect to the Sun, which is greatest when the Moon's apogee is in the octants of the Sun, and vanishes when the apogee arrives at the quadratures or syzygies ; and it is added to the mean motion while the Moon's apogee is passing from the quadrature of the Sun to the syzygy, and subtracted while the apogee is passing from the syzygy to the quadrature. Then, *Haec aequatio, quam semestrem vocabo, in octantibus apogaei, quando maxima est, ascendit ad 3'.45" circiter, quantum ex phaenomenis colligere potui. Haec est ejus quantitas in mediocri solis distantia a terra. Augetur vero ac diminuitur in triplicata ratione distantiae solo inverse, ideoque in maxima solis distantia est 3'.34", & in minima 3'.56" quamproxime : ubi vero apogaeum lunae situm est extra octantes, evadit minor ; estque ad aequationem maximam, ut sinus duplae distantiae apogaei lunae a proxima syzygia vel quadratura ad radium* (461/660).

By comparing the definition of Second Average with Newton's expression for his new concept of the semiannual equation, I have been definitely convinced that the concept of Second Average is directly translated and explained from Newton's *Principia*.

Third Average

In the section of " It is required to find Third Average " of *Yueli shuli*, it is said that : " The equations of the mean motion of the Moon are equal when the Sun are in the positions of the syzygies (*liangjiao*) and the quadratures (*daju*, elongation). It is subtracted from the mean motion when the Sun moves behind the syzygies, and added to it when the Sun moves behind the quadratures, and the greatest equation is 47", which is called Third Average, because of the North Pole is the pole of the circle described by the nodes, the parameter of the double distance of the Moon from the Sun is emphasised in *Xinfa suanshu*, i.e., *Xiyang xinfa lishu*, and the parameter of the double distance of the Sun from the node has been emphasized from the time of Newton... And the third average of every degree, is as the radius to the sine of double the distance of the Sun from the nodes "[10].

Newton round the " another equation of the mean motion of the Moon " by the same theory of gravity, and he pointed out that : *Et inde oritur alia medii motus lunaris aequatio, quam semestrem secundam vocabo, quaeque maxima est ubi nodi in solis octantibus versantur, & evanescit ubi sunt in syzygiis vel quadratures, & in aliis nodorum positionibus proportionalis est sinui duplae distantiae nodi alterutrius a proxima syzygia aut quadratura : additur vero medio motui lunae, si sol distat a nodo sibi proximo in antecedentia, subduci-*

---

10. *Yueli shuli*, " Qiu san pingjun ", in *ibid*.

tur si in consequentia, & in octantibus, ubi maxima est, ascendit ad 47" in mediocri solis distantia a terra, uti ex theoria gravitatis colligi (461/660).

There was a change from the second edition of Newton's *Principia* to the third edition, which was related to the original source for the Chinese translation in *Yueli shuli*. In the second edition, Newton said : *additur vero medio motui lunae, si sol distat a nodo sibi proximo in consequentia : dum Nodi transeunt a Solis Quadraturis ad proximas Syzygias, & subducitur in eorum transitu a Syzygiis ad Quadraturas* (461/660).

On the mean motion of the nodes of the Moon (Node Average)

In the section of " On the mean motion of the nodes of the Moon and the inclination of the Moon's orbit to the ecliptic plane (*lun jiaojun yu huang bai daju*) " in *Yueli shuli*, it is said that : " The velocity of the nodes of the Moon is changing as the inclination of the Moon's orbit to the ecliptic plane is variable, and the cause for this change was explained in detail in the previous book (i.e., *Lixiang kaocheng).* In Shoushi Calendar, which was derived from ancient method, the inclination of the Moon's orbit to the ecliptic plane was constantly $6°$ and without changing when in syzygies (*shuowang*) or quadratures (*liangxian*). So there was no change for the motion of the nodes of the Moon. In *Xinfa suanshu*, it were observed that the minimum of the inclination was $4°58'30"$ when the Moon is in syzygies, the maximum was $5°17'30"$ in quadratures, and the difference was $19'$. The maximum of the equation of the motion of the nodes was $1°4'68"$. From the time of Newton and Cassini, it has been said that the maximum of the inclination was $5°17'20"$ when the Sun is in syzygies (*liangjiao*), the minimum was $4°59'35"$ in quadratures (*ri jujiao jiushi du*), and the difference is $17'45"$. There is something added to the inclination when the Moon moves behind the syzygies : because of the increasing of the distances of the Sun from the Nodes and of the Moon from the Sun, the change of the equation of the inclination changes from the least to the maximum and comes to $2'43"$ when the distance of the Sun from the Nodes and of the Moon from the Sun are 90 degrees. The maximum of the equation of the mean motion of the Moon's nodes is $1°29'42"$. All of these is different from what were described in *Xinfa suanshu* "[11].

In *Principia*, Newton demonstrated at first that · *At si nodi in quadraturis consistunt, inclinatio minor est ubi luna versatur in syzygiis, quam ubi ea versatur in quadraturis, excessu 2'.43" ; uti in propositionis superioris Corollario quarto indicavimus.* (i.e., Corol. 4 of " Propositio XXXIV. Problema XV " of Book III. In the manuscript and 1st edition of *Principia*, Newton's value was 164", i.e., $2'44"$) (459/657).

___

11. *Yueli shuli*, " Qiu jiaojun ji huang bai daju ", in *ibid.*

Then, Newton required that the greatest variation of the inclination, abstracting from the situation of the Moon in its orbit, is 16'23 1/2". Therefore, the whole mean variation becomes 15'2", when the Moon is in the quadratures : and, increased by the same (the half of 2'43"), becomes 17'45" when the Moon is in the syzygies.

Si luna igitur in syzygiis constituatur, variatio tota in transitu nodorum a quadraturis ad syzygias erit 17'.45" : ideoque si inclinatio ubi nodi in syzygiis versantur, sit 5gr.17'.20" ; eadem, ubi nodi sunt in quadraturis, & luna in syzygiis, erit 4gr.59'.35". Atque haec ita se habere confirmatur ex observationibus (459/657).

Newton further gives that the equation of the motion of the nodes in the octants is 1°30'.

In *Principia,* Newton has revealed all the seven inequalities and perturbations of the motion of the Moon by his theory of gravity, which are : (1) the annual equation, corresponding to First Average in *Lixiang kaocheng houbian ;* (2) the position of the lunar apogee relative to the Sun, Newton's concept of the semiannual equation is corresponding to Second Average ; (3) the relation of the lunar nodes to the Sun, Newton's concept of the second semiannual equation is corresponding to Third Average ; (4) equation of centre, Newton's value for the greatest equation of the Sun's centre is 1°56'20", corresponding to the eccentricity of the Sun 16 11/12 ; (5) variation, Newton has calculated that : *Ideoque in apogaeo solis, variatio maxima est 33'.14", & in ejus perigaeo 37'.11", si modo eccentricitas solis sit ad orbis magni semi-diametrum transversam ut 16 15/16 ad 1000* (436/629).

The concept of " Second Even " (*er junshu*) has made use of this value. Newton has also calculated " the second equation of the Moon's centre ", the greatest of which comes to 2'25", and this value is made use of in *Yueli shuli* for " Third Even " (*san junshu*). (6) the sum of the distance of the Moon from the Sun, and of the Moon's Apoge from the Sun's Apoge is the Argument of the sixth Equation. Newton has pointed out that : *Et ut radius est ad sinum anguli sic inventi, ita 2'25" sunt ad aequationem centri secundam, addendam, si summa illa sit minor semicirculo, subducendam si major. Sic habebitur ejus longitudine in ipsis luminarium syzygiis* (463/662).

Correspondingly, there is " the Third Even of each degree, is as the radius to the sine of the angle " in *Lixiang kaocheng houbian.* (7). Say also " As the Radius to the Sine of the distance of the Moon from the Sun so is 1'30"±31 to the seventh equation ". It deserves the attention that the seventh lunar equation disappeared in the 3rd edition of *Principia,* leaving only six equations[12]. Correspondingly, there is " the last equation (*mojun*) comes to 1'30" when the dis-

---

12. I. Newton, *The Correspondence of Isaac Newton* (vol. 4, 1694-1709), Edited by J.F. Scott, Cambridge University Press, 1967, 322-327. Dr. N. Kollerstrom calls my attention to this problem by personal correspondence.

tance of the apogee of the Moon from the aphelion of the Sun is 90 degrees and the distance of the Moon from the syzygies of the Sun is 30 degrees... Therefore, Third Even for each degree of the distance of the Moon from the Sun, is as the radius to the sine of the distance of the Moon from the Sun ". It has been granted that the items 1, 4 and 5 above mentioned were known to contemporary astronomers, the remaining items originated with Newton[13].

It has been reached in the conclusion that *Sequel to the Compendium of Observational and Computational Astronomy*, which was published in 1742 and had been in practice in China from then to 1911, not only was translated and explained from Newton's theory of the Moon's motion, but also had directly taken advantage of the main data and calculated methods which were invented by Newton in his *Principia*. The astronomical theory for Guimao yuan Calendar was Newton's theory of the Moon's motion, furthermore the theory of gravity which was originated with Newton[14].

THE PROPAGATION OF NEWTON'S THEORY OF THE MOON'S MOTION IN CHINA

In 1687, Newton's theory of the Moon's motion, though being incomplets, was first issued in his *Principia*. In 27 February 1699/1700, Newton wrote a manuscript on the seven inequalities and perturbations of the Moon's motion, the main content of which, in the title of *Lunae Theoria Newtoniana*, was first published by David Gregory in his Latin edition of *Astronomiae Physicae et Geometricae Elementa* (1702)[15], and in English version in the same year[16]. As the result, in the history of the development of Newton's theory of the Moon's motion, there were two branches : one was from David Gregory's version, which is historically called *The Theory of the Moon's Motion*, the other was the change of Newton oneself expression on his theory of the Moon's motion in three Latin editions of his *Principia*. These two branches were in different degree relative to the propagation of Newton's theory of the Moon's motion in China.

In 1704, John Harris (*c*. 1666-1719) edited and included Newton's theory of the Moon's motion, under the heading of *Moon,* in his *Lexicon Technicum*[17].

13. D. Gjertsan, *The Newton Handbook,* N. Y., London, Routledge & Kegan Paul, 1986, 571-5/2.

14. E.W. Brown, Lu Jinggui (transl.), *Yueli chubian,* Tianjin, Tianjin baicheng shuju, 1936.

15. D. Gregory, *Astronomiae Physicae et Geometricae Elementa,* Oxford, 1702 ; London, 1715, a two-volume translation, Oxford, 1726.

16. I. Newton, *A New and most Accute Theory of the Moon's Motion. Whereby all her Irregularities may be solved, and her Place truly calculated to Two Minutes,* Written by that Imcomparable Mathematician Mr. Isaac Newton, London, Printed and sold by A. Baldwin in Warwick-lane, 1702.

17. J. Harris, *Lexicon Technicum, or An universal English Dictionary of Arts and Sciences : Explaining not only the Terms of Arts, but the Arts themselves,* London, 1704, 1708, 1716, 1725, 1736 ; vol. 2, 1710.

In 1707, Whilliam Whiston reprinted for the first time in his Cambridge lectures, *Praelectiones astronomicae*[18], which was the copy of the Latin version of 1702 . As to the variant versions of *The Theory of the Moon's motion*, there are Gregory (1726, Catalogue Numbers 1750, 1751 of the same version in Pe-T'ang, one of the Catholic missions in Beijing), Harris (1716, Catalogue n° 4077) and Whiston (1707, Catalogue n° 3088)[19]. It is worth noting that there is 123 tables in Whiston's *Praelectiones astronomicae* (1707, 328-450, and the tables in pages 451-459 are relative to the explanation of the calculation methods), in which 8 of " Astronomical solar tables " and 15 of " Lunar tables ". These 123 tables were translated into English and included in the volume 2 of Harris' *Lexicon Technicum* (1710, Catalogue n° 4077). It is possible that the astronomical solar and lunar tables in *Lixiang kaocheng houbian*, as the results of the observation and calculation of Chinese astronomers, were tabulated by Koegler and Pereira after they had referred to Whiston's *Praelectiones Astronomicae*, which was proved by Rong Zhenhua's expression : " Koegler has translated Newton's tables into Chinese ", although it seems that Newton hardly has tabulated so many tables[20].

More importantly, there is one copy of Newton's magisterial *Philosophiae Naturalis Principia Mathematica*, the 3rd Latin edition of 1726, in Pe-T'ang (Catalogue n° 2324). Having compared the data in *Lixiang kaocheng houbian* with Newton's data in there editions of the *Principia* and *The Theory of the Moon's Motion*, such as the values of 33'14" and 37'11" for Second Even and Newton's *variatio maxima*, respectively for the apogee and perigee of the Moon in the 2nd and the 3rd editions of *Principia,* were different from 33'40" and 37'25" in *The Theory of the Moon's Motion* ; the greatest of Third Even and Newton's *aequatio centri secunda* 2'25" was different from 2'10" in *The Theory of the Moon's Motion*[21], as well as the respective illustrations of the astronomical theory, we conclude that Koegler and Pereira had directly translated and taken advantage of Newton's expression of *The theory of the Moon's Motion* in the 2rd edition of *Principia* (1713)[22].

The area law and elliptic law (Fig. 1), which were demonstrated by Newton in Book I of *Principia*[23] and " Propositio LXVI. Theoria XXVI " of Book I (Fig. 2), which dealt with the problem of the interaction among the Sun, the Earth

18. W. Whiston, *Praelectiones Astronomicae,* Cambridge, 1707 ; London, 1715, English translation ; 1728, second issue of 1715.

19. Mission Catholique des Lazaristes à Pékin, *Catalogue de la Bibliothèque du Pe-T'ang,* Pékin, Imprimerie des Lazaristes, 1949.

20. Rong Zhenhua, Geng Sheng (transl.), *Zai Hua Yesu hui shilie zhuan ji shumu bubian,* vol. 1, Beijing, Zhonghua shuju, 1995, 336.

21. I. Newton, *Isaac Newton's Theory of the Moon's Motion (1702) : with a Bibliographical and historical introduction by I.B. Cohen,* Kent Dawson & Sons Ltd., 1975, 61.

22. *Yueli shuli,* " Qiu jiaojun ji huang bai daju ", in *ibid.*

23. C. Wilson, " From Kepler's Laws, So-called, to Universal Gravitation : Empirical actors ", *Archive for History of exact Sciences,* 6 (1970), 89-170.

and the Moon, were misinterpreted by missionaries as : " Recently, Western men Kepler, Cassini and others have observed and calculated, and regarded the orbital curve as elliptic which area was equally divided in the name of the average velocity (*pingxing du*). All of these is absolutely different from the past methods. But the values for the progression and retrogression of the Moon, which are derived from new method, are nearly equal to what were derived from the method of epicycle (*benlun*) and deferent (*junlun*), because of that the principles for the new method, though its coincidence is different from Tycho's, is the theory of *bentian gaobei* "[24].

Just because this misinterpretation, the scientific ideas in Newton's theory of the Moon's motion, which was derived from Newton's inverse square law and law of universal gravitation, were obscured. But having been viewed from another aspect, the transportation of ancient Chinese astronomical observation to the West in 1720s[25], the interpretation of Newton's *Principia* into Chinese calendar and the explanation of John Flamsteed (1664-1719)'s *Historia Coelestis Britannica* (1725, Catalogue n° 1610) and *Atlas* (1729) in *Yixiang kaocheng*, 1752, *Compendium to the Celestial Observational Astronomy*) have proved that the first half of eighteenth century is another period of great prosperity in the history of the exchange between East and West.

24. Richan shuli, " Richan zonglun ", in *Lixiang kaocheng houbian*, 1, *op. cit.*

25. S. Schaffer, " Halley, Delisle, and the making of the Comet ", *Standing on the Shoulders of Giants : A longer view of Newton and Halley,* University of Califomia Press, 1990, 254-298.

FIGURES

1. It is required to find the mean motion by the elliptic area
(in *Lixiang kaocheng houbian*, 1)

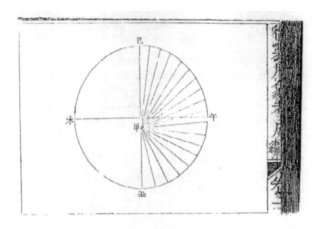

2. Newton's illustrative figure for Three-bodies Problem
" Propositio LXVI. Theoria XXVI " in *Principia* (1726), 171.

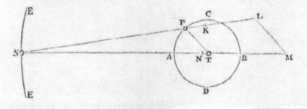

Propositio LXVI   279

corporum *T* & *P*. Hac vi fola corpus *P* circum corpus *T*, five immo-
tum, five hac attractione agitatum, defcribere deberet & areas, radio 10
PT, temporibus proportionales, & ellipfin cui umbilicus eft in centro
corporis *T*. Patet hoc per prop. xi. & corollaria 2. & 3. theor. xxi.
Vis altera eft attractionis *L M*, quæ quoniam tendit a *P* ad *T*, fu-

benlun 本輪
benri taiyang 本日太陽
bentian gaobei 本天高卑
Chen Zungui 陳遵媯
Chong zhen lishu 崇禎曆書
chujun 初均
daju 大距
Du Shiran 杜石然
er junshu 二均數
er pingjun 二平均
Geng Sheng 耿升
Gu Cong 顧琮
Guimao yuan Calendar 癸卯元曆
He Guozong 何國宗
Jiazi yuan Calendar 甲子元曆
jiaojun 交均
jindi dian 近地點
jinri dian 近日點
junlun 均輪
Kepler, Johan, Kaipuqin 開普勒
ou Ke bai'er 刻白爾
Koegler, Ignatius 戴進賢
lili 曆理
Lixiang kaocheng houbian 曆象考成後編
liangjiao 兩交
liangxian 兩弦
Lu Jinggui 盧景貴
« Lun jiaojun yu huang bai daju » 論交均與黃白大距
Mei Juecheng 梅瑴成
Ming Antu 明安圖
mojun 末均
Nai Duan 奈端
Pereira, André 徐懋德
pingxing du 平行度
« Qiu jiaojun ji huang bai daju » 求交均及黃白大距
ri jujiao jiushi du 日距交九十度
Rong Zhenhua 榮振華
san junshu 三均數

san pingjun 三平均
Schall von Bell, John Adam, Tang Ruowang 湯若忘
Shixian Calendar 時憲曆
shuowang 朔望
taiyang zhongju junshu 太陽中距均數
taiyin pingxing 太陽平行
taiyin yi pingjun 太陽一平均
Xiyang xinfa lishu 西洋新法曆書
Xinfa suanshu 新法算書
Xu Guangqi 徐光啓
yi pingjun 一平均
Yixiang kaocheng 儀象考成
yuandi dian 遠地點
yuanri dian 遠日點
Yueli chubian 月理初編
Yueli shuli 月離數理
« Yueli zonglun » 月理總論
yue tian gaobei zhongju hou sishiwu du 月天高卑中距四十五度
Zai Hua Yesu hui shilie zhuan ji shumu bubian 在華耶穌會士列傳及書目補編
zhengjiao pingxing 正交平行
zhengjiao yi pingjun 正交一平均
Zhonguo kexue jishu shigao 中國科學技術史稿
Zhongguo tianwen xue shi 中國天文學史
zui gaojun 最高均
zuigao pingxing 最高平行
zuigao yi pingjun 最高一平均

DOMINICAN CONTRIBUTIONS TO SCIENCE IN THE 16th AND 17th CENTURIES. THE EXAMPLE OF FRAY JUAN COBO IN EAST ASIA

José Antonio CERVERA JIMÉNEZ

The work of the Jesuits after they had already established themselves in China, in 17th and 18th centuries, is well-known. In fact, they were very important in the cultural and scientific relations between East and West during that period. But there were other missionaries, apart from Jesuits, who went to China (or tried to do it) and who made scientific work. I will approach their history, and I will try to show that they had also an important role in the relations between Europe and Asia.

In the 16th century, Europeans had already arrived to the Far East. At that time, Spain and Portugal were the strongest colonial powers, and they had fixed their limits of colonization in the Treaty of Tordesillas (1494). Generally speaking, we can say that the Portuguese had chosen their way to Asia towards the East, circumnavigating Africa and arriving to India, and the Spanish chose the way towards the West, arriving to America.

Both Portuguese and Spanish were willing to enter China. The Spaniards arrived to the Philippines as an intermediate point, but their objective was to establish themselves in the continent ; finally, they did not succeed, and they stayed in the Philippines. the Portuguese established themselves in Macao.

Generally speaking, the Jesuits went to China following the Portuguese way, from Europe to India and from there to Malacca and Macao ; the *friars* (Augustinians, Franciscans and Dominicans), many of whom were Spanish, went from Spain to Mexico and from there to the Philippines, from where they tried to enter China.

The first Jesuits who arrived to China, such as Michele Ruggieri or Matteo Ricci, entered from Macao, where they studied Chinese. In 1583, Ricci established himself in Guangdong and, finally, he arrived to Beijing in 1601, with the Spaniard Diego Pantoja. Ricci never left Beijing until his death in 1610. He was the first, but after him many Jesuits arrived in China. As we know very well, they were very important in the spread of European Science in China.

As regards the Spanish way, in 1565 the explorer Legazpi arrived to the Philippines. The Augustinians, who arrived with him, were the first missionaries in these Islands. One of them, Fray Andrés de Urdaneta (1498-1568), was an important explorer and cosmographer. Some years later, the Augustinians Fray Martín de Rada and Fray Jerónimo Marín went to China ; they were the first Spaniards who entered that country. Fray Martín de Rada (1533-1578) was very interested in science and was an important cosmographer. He studied Chinese and wrote *Arte y Vocabulario de la Lengua Chinense* (Art and Vocabulary of Chinese Language). Rada made astronomic observations and computations and studied the problems of longitude and the demarcation line between Spanish and Portuguese territories. He wrote to Europe in order to ask for the shipping out of some scientific books lost during the long journey. In a letter, he says : " of geometry I have here only Euclid and Archimedes ; of astronomy, Ptolemy and Copernicus... " (Cummins [1978, vol. 38, p. 81])[1]. González de Mendoza [1990, p. 360] says about Rada that he was an important geometer and mathematician. He wrote *Descripción del imperio de la Gran China* (Description of the Great China Empire) and *Política y Riquezas de la China* (Politics and Wealth of China), both of which were later used by Fray Juan González de Mendoza (1545-1618) in his book *Historia del Gran Reino de la China* (History of the Great Kingdom of China). This book has two parts. The first one deals with geography, history, culture, law, religions and traditions of China, and the second part consists of three travel stories. *Historia del Gran Reino de la China* was a *best-seller* in his time ; it was published in 1585 in Spanish, but by 1600, there were more than 10 Spanish editions and it had been translated into Italian, French, English, German, Dutch and Latin.

The Franciscans arrived to the Philippines in 1578. After them, the Jesuits, and after the Dominicans, arrived to Manila, the latter did it in 1587, from Mexico. The same year, three Dominicans travelled directly to Macao, also from Mexico, but they could not enter China, as they had wished. This was the first of nine attempts that the Dominicans made in order to establish themselves in China from the Philippines ; finally, Fray Angel Cocchi achieved their aim in 1632, founding a permanent mission in the continent.

There were some different reasons that explain the difficulty of friars to establish themselves in China. First, there was an intense rivalry between Spain and Portugal, and also a fierce competition between the Jesuits and the other missionaries. From the very beginning, China was a sort of monopoly for the Jesuits, and when Dominicans and Franciscans entered China, they found many difficulties. But the biggest problem, probably, was the controversy about Chinese rites. It was not clear whether the ceremonies to the ancestors and to Confucius performed by the Chinese were of a religious or only cultural

---

1. It is interesting to note that, as early as 1575, Copernicus was known by ecclesiastic intellectuals in Europe.

nature. The question was : Should a Christian Chinese be allowed to perform these ceremonies ? Matteo Ricci studied the problem and he concluded that they were not religious, and thereover they were allowed to Christian Chinese people. Most Jesuits (but not all) followed Ricci, and most Dominicans and Franciscans did not.

It is usually considered that Christianism was not successful in China due to the wrong attitude of Dominicans and Franciscans[2]. But this is not evident, and it is necessary to understand both positions. Some authors think that, even without the problem of Chinese rites, Christianism was not able to succeed in China because cultural differences are too wide[3]. It is possible that Christianism in China would have evolved to an eclecticism, a Christian religion with Chinese elements (Budist, Taoist, Confucianist), but this would not have been real Christianism, but some sort of heresy, a danger against which the friars fought.

On the relation of missionaries with Science, it is not true that only Jesuits were interested in Science, as it will be shown, and it is not true that the friars were not interested in Chinese culture, traditions and classical Chinese books. It is not true either that they did not learn the languages of the places where they went. Already in the 13th century, the Franciscans in China celebrated Mass in Mongolian, during the Yuan dynasty (Cummins [1978, vol. 38, p. 82]). In the 16th century, the missionaries in charge of spreading religion to Chinese people had to learn Chinese very hard ; the problem is that, since they could not enter China, they had to learn the language in Manila, where the pronunciation was not that of the Beijing dialect ; but, of course, they were able to read Chinese books.

As regards the scientific work of the Dominicans in the Far East, there were some important contributors, for example Fray Domingo Coronado (1615-1665), who wrote a book about astronomy ; Fray Ignacio Muñoz (1612-1685), who was a mathematician and an oceanographer ; and Fray Raimundo del Valle (1613-1683), who wrote a book about physiology. But maybe the most important is Fray Juan Cobo, who will be approached in this communication.

Fray Juan Cobo

He was born in Consuegra (Toledo province, Spain), in 1546 or 1547. He studied in Ocaña, Avila and Alcalá, and became a teacher at the monastery of

---

2. This is the point of view in many Histories written by Jesuits. See, for example, Bangert [1981], where the friars are described as *por lo general, poco entusiastas de introducir el cristianismo por medio de relojes y prismas, op. cit.*, 304 (in general, they did not like to introduce the Christianism through prisms and clocks).

3. See Garreau, 1983, vol. 8, 19-27. For the controversy of Chinese rites from a not-Jesuit point of view, see Cummins, 1978, vol. 38, 33-108.

Santo Tomás in Avila. He left Spain in 1587, stayed in Mexico some months and finally arrived to Manila in May 1588. When he arrived, he was assigned to the spread of religion to the Chinese, with whom Fray Miguel de Benavides was already working. Four years later, Toyotomi Hideyoshi (also known as Kwampaku-dono or Taiko-sama, who was the most powerful man in Japan at that time) sent an emissary to the Philippines. Then, the Spanish Governor, Pedro Gómez Dasmariñas, sent Cobo as an ambassador, in order to speak to Hideyoshi. Cobo went to Japan and he was successful in his mission. But when he was returning to the Philippines, his boat found a big storm and Cobo died.

It seems that he was very intelligent and therefore he learnt Chinese very quickly. As Aduarte [1962, vol. 1, p. 218] says : *él fue el primero que pública-mente predicó a los chinos, a cuyo sermón, como a cosa nunca vista y muy deseada, se halló presente el Gobernador de Manila, Santiago de Vera, con todo lo bueno de la ciudad, con no pequeña admiración suya y mucho mayor de los chinos, que nunca se persuadían a que una persona de otra nación pud-iese llegar a alcanzar tanto de su lengua* (he was the first that preached to the Chinese ; the Governor of Manila, Santiago Vera, was present during his ser-mon, and he was amazed to hear him as was the whole city, and specially the Chinese, because it was something never seen and very desirable to see that a foreign person could speak so well their language).

The life and the works of Cobo are a little similar to those of Matteo Ricci. Like Ricci, he was delighted when he learnt about Chinese culture. He could not expect such high culture in a non-European country. Bishop Salazar, in a letter that he sent to the King about the work of Cobo, wrote that it would be useless to try to invade China, and that the only way to reach Chinese people was by using intelligence, *pues se ve claro que con tal gente como ésta más ha de poder la fuerza de la razón que la de las armas* (Villarroel [1986, p. 15]) (because it is clear that with people like this, the power of reason is more use-ful that the power of weapons).

We can see here what Cobo would try to do with the *Shi Lu*, which is a more philosophical than religious book. He would try to use reason, to make the existence of God believable to Chinese intellectuals ; and, in order to achieve this purpose, he would use the study of God's creation : nature.

WRITTINGS BY FRAY JUAN COBO

- *Carta de la China* (Letter about China), where he writes about the Philip-pines and China, specially about the Chinese who live in Manila.

- *Doctrina Cristiana en Letra y Lengua china* (Christian Doctrine in Chi-nese language), where he explains in Chinese some religious questions. This is one of the first three books published in the Philippines.

- *Beng Sim Po Cam*[4], or, using pin yin transcription, *Ming Xin Shi Jian*, *Espejo Rico del Claro Corazón* (Rich Mirror of the Clear Heart), as Cobo translated. This is a bilingual book, handwritten ; in the left pages, we can see the original in Chinese, and in the right pages, the translation into Spanish by Cobo. The original is by a Chinese, Lip-Pun Luan, and it contains many moral sentences of the classical philosophers (Confucius, Mencius, *etc.*). The Dominicans were surprised to find a such high moral knowledge in a non-Christian culture. This book is very important, because it is the first Chinese book translated to a Western language.

- *Shi Lu.*

There are some other written documents, which could have been written by Cobo :

- *Vocabulario Chino* (Chinese Vocabulary). It is a study of the Chinese characters, with a method.

- *Arte de la Lengua China*. It is one of the first written foreign-Chinese Grammars.

- *Tratado de Astronomía* (Treatise of Astronomy). J.M. González. [1955-1967, vol. 5, p. 387] writes about a Treatise of Astronomy by Cobo. But, probably, he is speaking about the *Shi Lu*. In any case, we know that Cobo had some knowledge on astronomy, because we have references of his teaching of astronomy to Chinese people (Aduarte [1962, vol. 1, p. 219]).

THE *SHI LU*

The *Pien Cheng-Chiao Chen-Ch'uan Shih Lu* (or, in pin yin transcription, *Bian Zheng Jiao Zhen Chuan Shi Lu*) is Cobo's most important book, and one of the most important ever written about the scientific relations between East and West. There is a translation into Spanish and English, made by the Spanish Dominican Fidel Villarroel [1986], which has been used in this communication. Villarroel translated the title as *Testimony of the True Religion*.

The *Shi Lu* is one of the first three books printed in the Philippines. The other two are *Doctrina Cristiana en Letra y Lengua China*, also by Cobo, and *Doctrina Cristiana en Lengua Española y Tagala*, also by the Dominicans. It is not clear which of these three was published first. The *Shi Lu* was published in 1593, as it appears in the first page of the book, but the date of edition of the other two books is not so clear, even though it seems that they were printed at about the same time (see Villarroel [1986, pp. 53-57]). The method of printing was xylographic[5].

4. This title uses a transcription with the pronuntiation of the characters in some dialect of the South of China.

5. Xylographic impression is made with a wooden table, that is totally carved except for the characters (in opposite way), that remain raised pattern.

But probably, the most important aspect of the *Shi Lu* is its relation with the History of Chinese Science. The *Shi Lu*, as far as we know, is the first Chinese book that introduces European Science to the Chinese cultural world, written some years before the scientifical works by Matteo Ricci. Only for this reason, we can consider this book as one of the most important in all the History of scientific relations between West and East. Why, then, it has been nearly unknown for many scholars ? Probably, the main reason is that it only had some diffusion among the Chinese community in Manila, not in mainland China, and the only copy we know has been kept for a long time in Spain. For this reason this book is not well-known among many sinologists and historians of Science.

What did Cobo wish to accomplish when he wrote the *Shi Lu* ? We can compare it with his book *Doctrina Cristiana en Letra y Lengua China*, dedicated to the new Christian Chinese ; his aim is to let them know the fundamental truths of Catholic religion. However, the *Shi Lu* is devoted to highly educated people, to the Chinese non-Christian intellectuals who want to have a detailed rational explanation about why they should believe in Christianism. Cobo tries to show that Catholicism is not a foreign religion, as most Chinese thought, but a universal religion.

The title page of the *Shi Lu* has disappeared from the original, and the existing volume starts with the first page of the text. The book is not finished, because Cobo wanted to continue it after his journey to Japan, but he could not do it, because he died before coming back to the Philippines. There were three people involved in the composition of the book : Fray Juan Cobo, a scribe and a carver. The main person was Cobo, but the scribe (who, as far as it can be guessed, was a Christian Chinese, though we don't know his name) rewrote the book, correcting mistakes and introducing quotation from the Classical books, as it was usual in the Chinese books at that time.

The *Shi Lu* is an amazing example of cultural union between East and West. The book is structured as a dialogue, conversations between Cobo and the Chinese, between an European who admires China, and some Chinese who want to know the Truth. In the book, there are many questions made by Chinese to Cobo and his answers ; many paragraphs begin with *The Father said*, or *One says*, or *The Father answers*. It was clear to him that, in order to show the religion to such intelligent people, it was not possible to use methods different from reason.

Which European was the first to publish a book in Chinese language ? It seems that Michele Ruggieri was the first who published in Chinese, a *Catecismo*, in 1584, but in fact he wrote it in Latin and some Chinese translated it into Chinese. But the subject is mainly religious, and Cobo's *Shi Lu* is more philosophical and scientific. Matteo Ricci printed his *Catecismo* in 1604. This book, as Cobo's, is more philosophical than religious, but it is quite posterior to Cobo's *Shi Lu*. It seems clear, therefore, that Cobo was the first to publish in the Philippines, the first to translate a Chinese book into a Western language, and the first to write a philosophical-scientific book in Chinese, with European scientific knowledge.

CONTENTS OF THE *SHI LU*

The *Shi Lu* has nine chapters. The first three chapters are philosophical-theological, and the other six are scientific. The first chapter, " Discussion on the proofs of the true religion ", introduces the main objective of the book, to provide proofs that make believable the Catholic religion, the only true one for him. In the book (Villarroel [1986, p. 136]), he writes : " Under the heavens there is only one truth, not two. The true doctrine has only one culture, not two ".

After this introduction, Cobo begins with his main purpose : he wants to show that an Infinite Being exists, God, the Creator of the World and Man[6]. The second chapter reads " On the existence of an Infinite being, principle of all things ", and the third is " Talking about infinite things ". In order to demonstrate the existence of such an Infinite Being, he uses proofs similar to these used by Thomas Aquinas. However, Cobo can not finally prove it, because human mind is limited, and man can not understand the Infinite. But man can admire the created things, and therefore admire the Creator, God. This is the reason why chapters four to nine are devoted to the structure of the Earth, Heaven and Living Beings.

The fourth chapter is " On matters of geography ", but it could also be translated as " About the nature of the Earth ". This is the most important chapter of the book from the scientific point of view. Here, Cobo tries to show the usual European cosmological ideas of that time. Possibly the most important point is his wish to demonstrate to the Chinese that the Earth is round. It is probably the first time that it is clearly written in Chinese literature that the Earth is round, giving reasons in order to prove it ; the *Shi Lu* is, then, one of the most important documents of the whole History of Chinese science.

As Needham points out [1959, vol. 3, p. 210-228], there were three important cosmological theories in China, the *Gai Tian* theory (Hemispherical Dome), also known as *Zhou Bi*, the *Hun Tian* school (Celestial Sphere) and the *Xuan Ye* theory (Infinite Empty Space). The traditional idea in China is to think that the Earth is flat and square, and that Heaven is spherical, as it appears in some texts on the *Gai Tian* theory. However, Needham, on analysing some texts, thought that the *Hun Tian* theory considers the Earth spherical. Nowadays, most researchers do not agree with this idea. For example, Cullen points out that there is no evidence in Chinese texts mentioning the sphericity of the

---

6. The word used by Cobo to refer to the Infinite is *Tai Ji*, very important in Chinese Philosophy. For a discussion about the history of this term in Chinese Philosophy, see Lundbaek [1983].

Earth. The Chinese did not need to consider the sphericity of the Earth to achieve an important development of Astronomy, with very accurate predictions, because they used computational procedures, not geometrical. As Cullen [1996, p. 39] says : " the differences between the so-called Hun Tian and Zhou Bi " schools " in ancient China can be most clearly understood in connection with changes in instrumentation and observational procedures ".

Cobo gives arguments in order to demonstrate the sphericity of the Earth in this fourth chapter, where there are some drawings that show some of these arguments. For example, one of the proofs says that it is only possible to see the mast of a boat in the sea, far away, and the nearer to the coast it is, the larger part of the boat can be seen. Maybe the most interesting proof is that during a lunar eclipse, we see the shadow of the Earth on the Moon ; since the shadow is round, the Earth should also be round.

The first drawing in this fourth chapter shows the system of the world at that time, the geocentrical Ptolemaic Universe, because, although the *Shi Lu* was printed nearly half a century after the *De Revolutionibus* of Copernicus, the usual ideas were still the former[7].

The fifth chapter is " About the reality of the earthly things ". Here, there are some physical ideas of that time. For example, he says : " The nature of the earth is to be dry ; that of the water is to be humid " (Villarroel [1986, p. 286]) ; this can remind us of Aristotle, whose ideas had not been surpassed at that time. In the chapter, Cobo discusses the nature and properties of the water, the earth and the air.

After Astronomy, Geography and Physics, Cobo begins to speak about the created things on the Earth, about the living beings. In the sixth chapter, " On the plants of the earth and other vegetables ", Cobo says that God has given to every plant its form and purpose. All the created was made for the use of man. In particular, plants have four possible uses : to provide beauty, to provide food, to provide materials (like wood) and to cure the body (as medicines). The seventh chapter is " On things of the animal kingdom ". Here, the Chinese ask Cobo why there are some dangerous or poisonous animals, if the animals were created for the utility of men. Cobo, in this point, has to talk about religion, and says that this is a punishment for the first sin of Adam ; this is one of the

7. As we know, the geocentrical Ptolemaic system had arrived to a great simplicity and accuracy in the 15th century with Peuerbach and Regiomontanus. For religious people in the 16th century, some of the most popular texts were A. Piccolomini's *Della Sfera del Mondo* (first edition 1540) and the Jesuit C. Clavius' *In Sphaeram Ioannis de Sacrobosco Commentarius* (Rome, 1585). This last title was the text that M. Ricci mainly used. However, Villarroel [1986, p. 87] points out that Cobo may have used Piccolomini's book, because the Dominicans carried it to the Philippines and an edition of 1564 (Venice) is still today in the Library of the University of Santo Tomás, in Manila. However, this is not very clear, since Piccolomini' system comprises 10 spheres, but the tenth is the *Primo Mobile*, a movible sphere, and therefore it does not correspond to Cobo's tenth sphere, which does not move.

few places in these chapters where he introduces some comments about religion.

The last chapters of the *Shi Lu* are curious and funny. The eighth chapter is " On how the animals know what they should eat and drink ", and the ninth " On how the animals of the world know the medicines they must take ". In these chapters there are also some drawings ; there we can see crabs, a fox, some birds, *etc*. In this point, the *Shi Lu* finishes. Cobo wanted to continue it, but he was not able to do it, because he died when he was going back from Japan to the Philippines.

BIBLIOGRAPHY

D. Aduarte, O.P., " Historia de la Provincia del Santo Rosario de la Orden de Predicadores en Filipinas, Japón y China ", Series " *Biblioteca Missionalia Hispanica* ", Series A, XIV, Madrid, Consejo Superior de Investigaciones Científicas, Departamento de Misionología Española, 1962, (Raycar S.A. Impresores), 2 vols. Reprint of Aduarte's book, published in Manila in 1640.

W.V. Bangert, S.I., *Historia de la Compañía de Jesús*, Santander, Sal Terrae, 1981. Translation of the English edition.

A.M. de Castro, O.S.A., " Misioneros Augustinos en el Extremo Oriente 1565-1780 ", Series " *Biblioteca Missionalia Hispanica* ", Series B, VI, Madrid, Consejo Superior de Investigaciones Científicas, Instituto de Santo Toribio de Mogrovejo, Ediciones Jura, 1954.

J. Cobo, O.P., *Bian Zheng Jiao Zhen Chuan Shi Lu*, Manila, 1593. The only copy known today is in the National Library of Madrid.

C. Cullen, " Astronomy and Mathematics in Ancient China : the Zhou Bi Suan Jing ", *Needham Research Institute Studies*, 1, Cambridge, Cambridge University Press, 1996.

J.S. Cummins, " Two Missionary Methods in China : Mendicants and Jesuits ", *Archivo Ibero-Americano (Spain)*, 38 (149-152), (1978), 33-108.

J. Garreau, " Chinese Reaction to Christianity ", *Asian Thought & Society*, 8 (22-23), (1983), 19-27.

D. González, O.P., " Irradiación Misionera de Santo Tomás de Avila ", included in the book *Monjes y Monasterios Españoles* (Proceeding of the Symposium, San Lorenzo del Escorial, 1/5-9-1995).

J.M. González, O.P., *Historia de las Misiones Dominicanas de China*, Madrid : Imprenta, Juan Bravo 3, 1955-1967, 5 vols.

J. González de Mendoza, O.S.A., " Historia del Gran Reino de la China ", Series " *Biblioteca de Viajeros Hispánicos* ", 6 (Madrid, Miraguano Ediciones and Ediciones Polifemo, 1990. Reproduction of the book published by González de Mendoza in 1585.

J. González Vallés, O.P., (ed), " Cuatro Siglos de Evangelización (1587-1987), Rutas Misioneras de los Dominicos de la Provincia de Nuestra Señora del Rosario ", *Series " Orientalia Dominicana - General ",* 2 (Madrid, Huellas Dominicanas, 1987).

K. Lundbaek, " The Image of Neo-Confucianism in *Confucius Sinarum Philosophus ", Journal of the History of Ideas,* 44 (1), (1983), 19-30.

J. Needham, *Science and Civilization in China,* Cambridge, Cambridge University Press, 1959, vol. 3.

H. Ocio, O.P., *Compendio de la Reseña Biográfica de los Religiosos de la Provincia del Santísimo Rosario de Filipinas,* Manila, Establecimiento Tipográfico del Real Colegio de Santo Tomás, 1895.

L. Pérez, O.F.M., " Origen de las Misiones Franciscanas en la Provincia de Kwuang-Tung (China) ", included in the *Archivo Ibero-Americano,* numbers XX-XXIII (Madrid, Imprenta de G. López del Horno, 1918).

L. Pérez, O.F.M., " Labor Patriótica de los Franciscanos en el Extremo Oriente ", included in the *Archivo Ibero-Americano,* numbers 94-96 (Madrid, Imprenta de los Hijos de Tomás Minuesa de los Ríos, 1929).

F. Villarroel, O.P., " Pien Cheng-Chiao Chen-Ch'uan Shih Lu. Apología de la Verdadera Religión ", *Series " Orientalia Dominicana - Filipinas ",* 3 (Manila, UST Press, 1986). Facsimile reproduction of the original Chinese text printed in Manila in 1593, with introduction by A. Santamaría, A. Domínguez and F. Villarroel, ed. by Villarroel.

CONTRIBUTORS

Luis Carlos ARBOLEDA
Grupo de Historia de las Matemáticas
Universidad del Valle
Cali (Colombia)

Ana CARNEIRO
CICSA, History of Science Unit
Faculdade de Ciências e Tecnologia
Universidade Nova de Lisboa
Monte de Caparica (Portugal)

José Antonio CERVERA JIMÉNEZ
Seminario de Historia de la Ciencia
Dep. de Matemática Aplicada
Universidad de Zaragoza
Zaragoza (Spain)

Maria Paula DIOGO
CICSA, History of Science Unit
Faculdade de Ciências e Tecnologia
Universidade Nova de Lisboa
Monte de Caparica (Portugal)

FUNG Kam-Wing
University of Hong-Kong
Department of Chinese
Hong Kong (P.R. China)

Carlos D. GALLES
Departamento de Fisica
Facultad de Humanidades y Artes
Universidad Nacional de Rosario
Rosario (Argentina)

Keizo HASHIMOTO
Faculty of Sociology
Kansai University
Osaka (Japan)

Arne HESSENBRUCH
Department of History and
Philosophy of Science
Cambridge (United Kingdom)

Antonio LAFUENTE
Centro de Estudios Históricos (CSIC)
Madrid (Spain)

Celina A. LÉRTORA MENDOZA
CONICET
Buenos Aires (Argentina)

LU Dalong
Institute for History of Natural Sciences
Chinese Academy of Sciences
Beijing (P.R. China)

Efthymios NICOLAÏDIS
The National Hellenic Research
Foundation
Athens (Greece)

A.A. PECHENKIN
Institute for History of Science and
Technology of Russian Academy of
Sciences
Moscow Physico-Technological Institution
Moscow (Russia)

María de la PAZ RAMOS LARA
Center for Interdisciplinary Research
in the Sciences and Humanities
National Autonomous University of
Mexico (Mexico)

Juan PIMENTEL
Centro de Estudios Históricos (CSIC)
Madrid (Spain)

Juan José SALDAÑA
Faculty of Philosophy and Letters
National Autonomous University of
Mexico (Mexico)

D.L. SAPRYKIN
Institute for History of Science and
Technology of Russian Academy of
Sciences
Moscow Physico-Technological Ins-
titution
Moscow (Russia)

Ana SIMÕES
Department of Physics
Universidade de Lisboa
Lisboa (Portugal)

Antonio E. TEN
University of Valencia
Valencia (Spain)

George N. VLAHAKIS
The National Hellenic Research
Foundation
Athens (Greece)

Christos T. XÉNAKIS
Université de Thessalie
Thessalie (Grèce)